Collins

AF775434

Greater Depth Maths

Build curiosity and confidence

TEACHER GUIDE KEY STAGE 1

Herts for Learning

William Collins' dream of knowledge for all began with the publication of his first book in 1819. A self-educated mill worker, he not only enriched millions of lives, but also founded a flourishing publishing house. Today, staying true to this spirit, Collins books are packed with inspiration, innovation and practical expertise. They place you at the centre of a world of possibility and give you exactly what you need to explore it.

Collins. Freedom to teach.

Published by Collins
An imprint of HarperCollins*Publishers*
The News Building
1 London Bridge Street
London
SE1 9GF

HarperCollins*Publishers*
1st Floor, Watermarque Building
Ringsend Road
Dublin 4
Ireland

Browse the complete Collins catalogue at

www.collins.co.uk

10 9 8 7 6 5 4 3 2 1

ISBN 978-0-00-845489-0

British Library Cataloguing-in-Publication Data
A catalogue record for this publication is available from the British Library.

Authors: Charlie Harber, Rachael Brown, Nicola Adams, Laura Dell
Publisher: Lizzie Catford
Product manager: Letitia Luff
Editorial management: Oriel Square
Cover designer: Amparo Barrera
Cover photograph: ADragan/Shutterstock
Typesetter: Jouve India Pvt. Ltd.
Internal illustrations: Jouve India Pvt. Ltd.
Production controller: Alhady Ali
Printed and bound in the UK using 100% Renewable Electricity at CPI Group (UK) Ltd

FSC www.fsc.org
MIX
Paper from responsible sources
FSC C007454

This book is produced from independently certified FSC™ paper to ensure responsible forest management.

For more information visit: www.harpercollins.co.uk/green

The publishers gratefully acknowledge the permission granted to reproduce the copyright material in this book. Every effort has been made to trace copyright holders and to obtain their permission for the use of copyright material. The publishers will gladly receive any information enabling them to rectify any error or omission at the first opportunity.

Digital downloads available at:

www.collins.co.uk/HertsforLearning/downloads

Contents

Getting Started Guide

What is greater depth, and what does it look like?

Greater-depth learning is more than cognitive content; it is complex and pays attention to mathematical structure and what lies beneath a concept or procedure. A child working at greater depth is ready and able to explore mathematical concepts in more detail. Key characteristics include being curious and asking interesting mathematical questions, making connections and identifying patterns, collaborating with others effectively, communicating their ideas with increasing clarity, being creative with their responses, and considering arguments and providing counter-arguments with proof.

Collins Greater Depth Maths

The **Collins Greater Depth Maths** resources have been written by an experienced team of maths practitioners. Features of these resources include:

- high-quality mathematical investigations that spark curiosity;
- professional development embedded to increase teacher confidence;
- ready-to-teach sequences;
- 'stumble support' to provide inclusive greater-depth opportunities for all children.

The resources are provided for three phases of the primary curriculum: Key Stage 1, Lower Key Stage 2 and Upper Key Stage 2. There are 15 tasks for each phase, all fully explained and supported.

Each task within a phase has a comprehensive set of resources. The **Teacher Guide** provides detailed guidance showing teachers how to teach and facilitate each task, with different options to suit classroom needs. Each task comprises a set of component elements and could take a full lesson or several lessons. Children are given extensive opportunities to explore the mathematical concept under investigation and to discuss and share ideas.

The innovative **Pupil Resources** contain task-specific photocopiable pages that support children at each stage of a task, to develop independence.

Key Stage 1 tasks and the National Curriculum

The charts on pages 5–8 list the KS1 tasks and their relevant National Curriculum objectives.

All tasks can be done in either Year 1 or Year 2, but to give some guidance, the chart suggests which Year group could focus on a given task. The allocation between Years 1 and 2 ensures good coverage of different maths topics across the Year groups. To provide teachers with some flexibility, three of the tasks are allocated to both Years 1 and 2.

In the **Teacher Guide** for each task, a small icon indicates the Year group for which the task is suggested.

Year 1 Year 2

<table>
<tr><th></th><th>Year 1 National Curriculum</th><th>Year 2 National Curriculum</th></tr>
<tr><td colspan="3">Number and place value</td></tr>
<tr><td rowspan="2">1 I Can Read Your Mind
Year 1</td><td>• Read and write numbers from 1 to 20 in numerals and words.</td><td>• Read and write numbers to at least 100 in numerals and in words.</td></tr>
<tr><td colspan="2">Links to White Rose: Year 1 Autumn – Number: Place Value (within 20)</td></tr>
<tr><td rowspan="2">2 Hidden Dragon Eggs
Year 1</td><td>• Identify and represent numbers using objects and pictorial representations including the number line, and use the language of: equal to, more than, less than (fewer), most, least.
• Pupils practise […] counting in twos, fives and tens from different multiples to develop their recognition of patterns in the number system (for example, odd and even numbers) […] [non-statutory guidance]</td><td>• Use place value and number facts to solve problems.</td></tr>
<tr><td colspan="2">Links to White Rose: Year 1 Autumn – Number: Place Value (within 20)</td></tr>
<tr><td rowspan="2">3 Dots and Dashes
Year 1</td><td>• Identify and represent numbers using objects and pictorial representations including the number line, and use the language of: equal to, more than, less than (fewer), most, least.
• Pupils recognise and create repeating patterns with objects and with shapes. [non-statutory guidance]</td><td>• Identify, represent and estimate numbers using different representations, including the number line.
• Using materials and a range of representations, pupils practise counting, reading, writing and comparing numbers to at least 100 and solving a variety of related problems to develop fluency. [non-statutory guidance]</td></tr>
<tr><td colspan="2">Links to White Rose: Year 1 Summer – Place Value (within 100)</td></tr>
<tr><td colspan="3">Addition and subtraction</td></tr>
<tr><td rowspan="2">4 Money Maze
Year 1 or 2</td><td>• Add and subtract one-digit and two-digit numbers to 20, including zero.
Measurement:
• Recognise and know the value of different denominations of coins and notes.</td><td>• Add and subtract numbers using concrete objects, pictorial representations, and mentally, including: a two-digit number and ones; a two-digit number and tens; two two-digit numbers; adding three one-digit numbers.
Measurement:
• Find different combinations of coins that equal the same amounts of money.
• Solve simple problems in a practical context involving addition and subtraction of money of the same unit, including giving change.</td></tr>
<tr><td colspan="2">Links to White Rose: Year 1 Summer – Measurement: Money
Year 2 Autumn – Number: Addition and Subtraction
Year 2 Autumn – Measurement: Money</td></tr>
</table>

	Year 1 National Curriculum	Year 2 National Curriculum
Addition and subtraction		
5 **Grid Patterns** Year 2	• Represent and use number bonds and related subtraction facts within 20. • Solve one-step problems that involve addition and subtraction, using concrete objects and pictorial representations, and missing number problems such as 7 = ? – 9.	• Solve problems with addition and subtraction: • using concrete objects and pictorial representations, including those involving numbers, quantities and measures; • applying their increasing knowledge of mental and written methods. • Recall and use addition and subtraction facts to 20 fluently, and derive and use related facts up to 100.
	Links to White Rose: Year 2 Autumn – Number: Addition and Subtraction	
Multiplication and division		
6 **All the Arrays** Year 1	• Solve one-step problems involving multiplication and division, by calculating the answer using concrete objects, pictorial representations and arrays with the support of the teacher.	• Solve problems involving multiplication and division, using materials, arrays, repeated addition, mental methods, and multiplication and division facts, including problems in contexts.
	Links to White Rose: Year 1 Summer – Number: Multiplication and Division	
7 **Crossing the River** Year 1 or 2	• Solve one-step problems involving multiplication and division, by calculating the answer using concrete objects, pictorial representations and arrays with the support of the teacher. **Number and place value:** • Count, read and write numbers to 100 in numerals; count in multiples of twos, fives and tens.	• Solve problems involving multiplication and division, using materials, arrays, repeated addition, mental methods, and multiplication and division facts, including problems in contexts. **Number and place value:** • Count in steps of 2, 3, and 5 from 0, and in tens from any number, forward and backward.
	Links to White Rose: Year 1 Summer – Number: Multiplication and Division Year 2 Spring – Number: Multiplication and Division	
8 **Golden Crowns** Year 2	• *Pupils make connections between arrays, number patterns, and counting in twos, fives and tens.* [non-statutory guidance]	• Recall and use multiplication and division facts for the 2, 5 and 10 multiplication tables, including recognising odd and even numbers. **Number and place value:** • Count in steps of 2, 3, and 5 from 0, and in tens from any number, forward and backward.
	Links to White Rose: Year 2 Spring – Number: Multiplication and Division	

	Year 1 National Curriculum	Year 2 National Curriculum
Fractions		
9 **Can I Halve It?** Year 1	• Recognise, find and name a half as one of two equal parts of an object, shape or quantity.	• *Pupils use fractions as 'fractions of' discrete and continuous quantities by solving problems using shapes, objects and quantities. They connect unit fractions to equal sharing and grouping, to numbers when they can be calculated, and to measures, finding fractions of lengths, quantities, sets of objects or shapes.* [non-statutory guidance]
	Links to White Rose: Year 1 Summer – Number: Fractions	
10 **Poppy the Penguin** Year 2	• Recognise, find and name a half as one of two equal parts of an object, shape or quantity. • Recognise, find and name a quarter as one of four equal parts of an object, shape or quantity.	• Recognise, find and name fractions $\frac{1}{3}$, $\frac{1}{4}$, $\frac{2}{4}$ and $\frac{3}{4}$ of a length, shape, set of objects or quantity. **Multiplication and division:** • Solve problems involving multiplication and division, using materials, arrays, repeated addition, mental methods, and multiplication and division facts, including problems in contexts.
	Links to White Rose: Year 2 Spring – Number: Fractions	
Measurement		
11 **What's in the Box?** Year 1 or 2	• Recognise and know the value of different denominations of coins and notes. **Addition and subtraction:** • Add and subtract one-digit and two-digit numbers to 20, including zero.	• Find different combinations of coins that equal the same amounts of money. • Solve simple problems in a practical context involving addition and subtraction of money of the same unit, including giving change. **Addition and subtraction:** • Add and subtract numbers using concrete objects, pictorial representations, and mentally, including: a two-digit number and ones; a two-digit number and tens; two two-digit numbers; adding three one-digit numbers.
	Links to White Rose: Year 1 Summer – Measurement: Money Year 2 Autumn – Measurement: Money	
12 **Built to Scale** Year 2	• Compare, describe and solve practical problems for: • lengths and heights [for example, long / short, longer / shorter, tall / short, double / half].	• Choose and use appropriate standard units to estimate and measure length / height in any direction (m / cm); mass (kg / g); temperature (°C); capacity (litres / ml); to the nearest appropriate unit, using rulers, scales, thermometers and measuring vessels. **Fractions:** • Recognise, find, name and write fractions $\frac{1}{3}$, $\frac{1}{4}$, $\frac{2}{4}$ and $\frac{3}{4}$ of a length, shape, set of objects or quantity. • Write simple fractions for example, $\frac{1}{2}$ of 6 = 3 and recognise the equivalence of $\frac{2}{4}$ and $\frac{1}{2}$.
	Links to White Rose: Year 2 Summer – Measurement: Length and Height	

	Year 1 National Curriculum	Year 2 National Curriculum
Geometry		
13 **Peg Board Creations** Year 1	**Geometry – properties of shapes:** • Recognise and name common 2D and 3D shapes, including: • 2D shapes [for example, rectangles (including squares), circles and triangles]. • *Pupils recognise these shapes in different orientations and sizes, and know that rectangles, triangles, cuboids and pyramids are not always similar to each other.* [non-statutory guidance] **Geometry – position and direction:** • *Pupils use the language of position, direction and motion, including: left and right, top, middle and bottom [...]* [non-statutory guidance]	**Geometry – properties of shapes:** • Compare and sort common 2D and 3D shapes and everyday objects. **Geometry – position and direction:** • Use mathematical vocabulary to describe position, direction and movement, including movement in a straight line and distinguishing between rotation as a turn and in terms of right angles for quarter, half and three-quarter turns (clockwise and anticlockwise).
	Links to White Rose: Year 1 Autumn – Geometry: Shape Year 1 Summer – Geometry: Position and Direction	
14 **Story Shapes** Year 2	**Geometry – properties of shapes:** • Recognise and name common 2D and 3D shapes, including: • 2D shapes [for example, rectangles (including squares), circles and triangles]. • *Pupils recognise these shapes in different orientations and sizes, and know that rectangles, triangles, cuboids and pyramids are not always similar to each other.* [non-statutory guidance]	**Geometry – properties of shapes:** • Identify and describe the properties of 2D shapes, including the number of sides and line symmetry in a vertical line. • Compare and sort common 2D and 3D shapes and everyday objects. • *Pupils identify, compare and sort shapes on the basis of their properties and use vocabulary precisely, such as sides, edges, vertices and faces.* [non-statutory guidance]
	Links to White Rose: Year 2 Spring – Geometry: Properties of Shape	
Statistics		
15 **Pattern People** Year 2	N/A	• Interpret and construct simple pictograms, tally charts, block diagrams and simple tables. • Ask and answer simple questions by counting the number of objects in each category and sorting the categories by quantity. • Ask and answer questions about totalling and comparing categorical data.
	Links to White Rose: Year 2 Spring – Statistics	

The components

There are three components for each task:

- **Teacher Guide**
- **Teacher's Presentation PowerPoint slides**
- **Pupil Resources.**

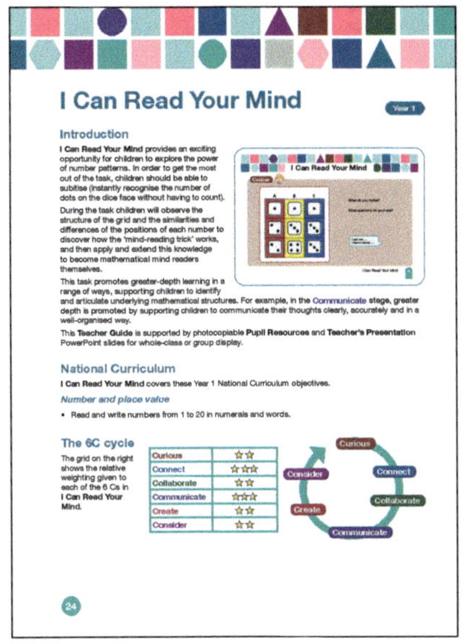

Teacher Guide

The **Teacher Guide** gives detailed teaching pathways for each task, including:

- guidance on key mathematical ideas;
- examples of age-appropriate children's responses;
- advice on how to model the identified strategies;
- prompts and sentence stems to scaffold children's learning and maintain engagement;
- word banks to deepen and focus vocabulary.

Teacher's Presentation PowerPoint slides

PowerPoint slides are provided, to enhance whole-class teaching and discussion. Some of the slides include notes providing 'stumble support' suggestions, that may help you to scaffold children in their problem-solving thinking and enable access for children who are less confident.

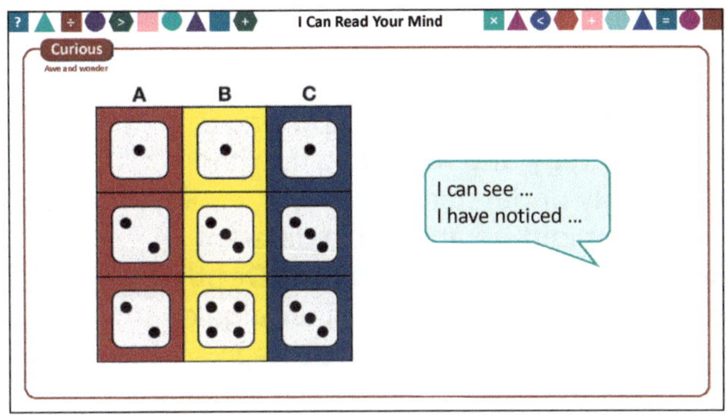

Where a printable handout might be useful, these are provided as black-and-white slides.

There are also slides of the **Pupil Resources**, should you wish to display these.

Pupil Resources

The **Pupil Resources** comprise printable pages for classroom distribution. The pages show visual representations of the problem being investigated, with prompts and word banks to help children discuss and develop their ideas. For each task there are four pages of **Pupil Resources:**

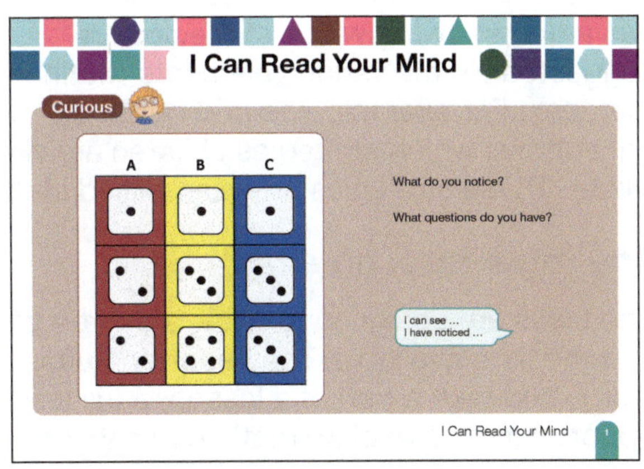

- For tasks allocated to either Year 2, or to both Year 1 and Year 2, there is one page each for the **Full** and the **Express Curious** tasks (see below for an explanation of **Full and Express Pathways**), and two pages for the remaining 5C tasks which are common to both pathways.
- For tasks allocated to Year 1 only, there is no **Express Pathway**, and instead the **Curious** task is spread across the first two pages of the **Pupil Resources**.

It is suggested that teachers photocopy the pages and laminate them to create a reusable card-like resource, giving children freedom and flexibility to explore the tasks.

The key features of the resource

The 6C cycle

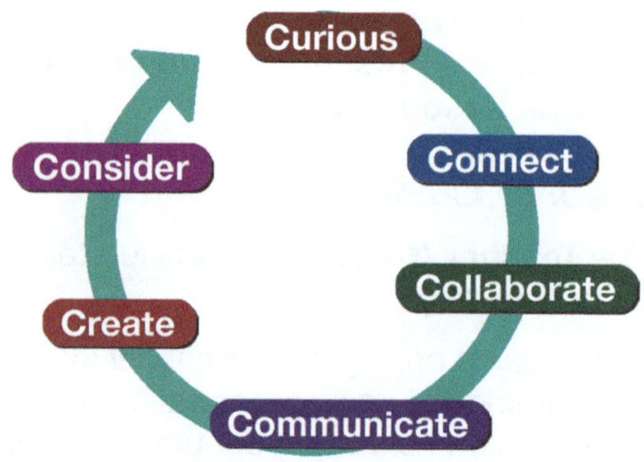

The 6Cs represent aspects of the desired characteristics of greater-depth learning: to be curious, to connect mathematical concepts, to collaborate with one another, to communicate their ideas and findings, to create their own examples and to consider and review their work and the work of others.

- **Curious:** promoting and encouraging curiosity in children.
- **Connect:** supporting children to connect their immediate observations to their past learning.
- **Collaborate:** developing meaningful collaboration between peers as a worthwhile skill, not just in mathematics but in everyday life.
- **Communicate:** supporting children to communicate mathematically with precision, clarity and respect.
- **Create:** fostering further curiosity and creativity.
- **Consider:** reflecting on mathematical learning experiences.

Problem-solving heuristics

Children are often good at applying a single strategy and repeating it, without necessarily understanding why it works, or how it might be applied in different situations. The tasks in **Greater Depth Maths** encourage the use of a range of problem-solving heuristics (strategies), and promote the discussion and investigation of the different types of strategy for different types of problem. Each task has a grid showing which strategies are suggested for the various elements within that task. For example, the grid on the right shows which strategies children are likely to use in their investigation of the task **Money Maze**. These strategies are exemplified in the **Connect** stage of the task.

Draw a diagram	
Make a table / spot a pattern	
Make a model / act it out	✓
Try then improve	✓
Create a systematic list	✓
Solve a simpler but related problem	✓
Work backwards	✓

The route through a task

The tasks are designed to allow a degree of flexibility, according to the confidence of the children and the time available. The **Teacher Guide** offers a route through each task, using the 6Cs. Each task could take a part of a lesson, a full lesson or several lessons. Children are given extensive opportunities to explore mathematical concepts and to discuss and share ideas. Some children will make links more quickly, and will be able to skip some of the steps.

Full and Express Pathways

For tasks allocated to Year 2, or to both Year 1 and Year 2, there are two main ways to use each task: the **Full 6C Pathway** and the **Express Pathway**. The Full Pathway exemplifies each stage of the journey through the 6Cs. However, there will be some occasions where there may not be sufficient opportunity to complete the Full Pathway, and teachers may choose to use the Express Pathway instead.

A Full Pathway could take a full lesson or several lessons, whereas an Express Pathway could form part of a lesson or could be returned to over a series of lessons.

Each task has two **Pupil Resource** pages for the **Curious** task: the first page shows the 'Full' version, and the second page shows a slightly modified 'Express' version. When following the **Full Pathway**, teachers should use the prompts in the Awe and wonder and Develop and deepen sections of the **Teacher Guide**, to encourage discussion and exploration, and allow children to spend a significant amount of time developing their own mathematical questions.

Although the **Express Pathway** prompt is similar, it is a little more direct and focused. A much shorter discussion would take place, before moving fairly quickly onto the **Connect** tasks. Teachers can then choose how much of the rest of the cycle the children engage in. Although the Express Pathway will involve fewer stages, the quality of the explanation or response should remain high. The expectation is that children would still explore the structure of the mathematics rather than just replicate the examples provided in an attempt to provide proof.

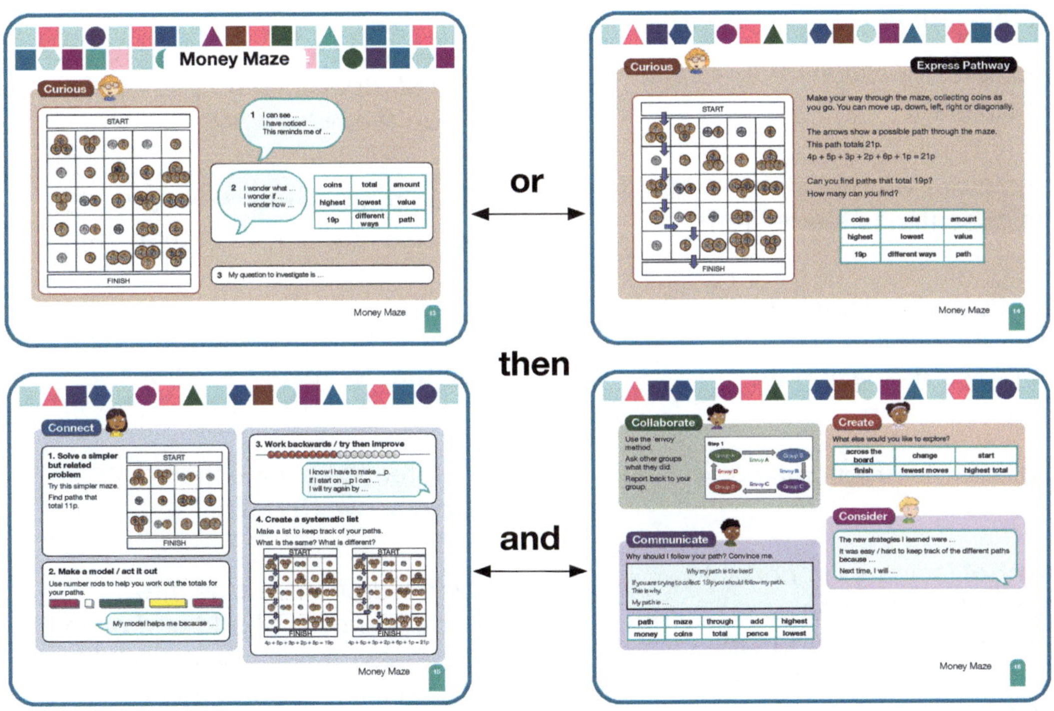

Each Pathway has its own **Curious Pupil Resource** page**.**

The tasks and the **Pupil Resource** pages for the remaining 5Cs are common to both pathways.

Stumble support

The in-built 'stumble support' provides inclusive greater-depth opportunities for all by enabling scaffold-enabled access. You will find explicit references to 'stumble support' within the notes section of some of the **Teacher's Presentation** slides, but stumble support is also woven throughout all the components, in the form of word banks, prompts and sentence stems.

The **How to use** sections below explain the different 6C elements within each task, briefly describing the purpose of each element, key planning questions and details of any additional scaffolds provided. It is useful to look at the **Teacher Guide** and the **Pupil Resources** together, so that decisions concerning when to distribute the **Pupil Resources** can be made. The **Pupil Resources** do not have to be shared immediately and in some cases, the **Curious** stage will give guidance on how to introduce the question / problem to the children *before* the **Pupil Resources** are shared.

The **Teacher's Presentation** slides provide flexibility in terms of display and also ongoing 'stumble support'.

How to use the Teacher Guide

The tasks work with a series of prompts from the teacher, that can be used as whole-class prompts, or with smaller groups. These include key structural questions and prompts such as:

- *What's the same? What's different?*
- *Which is harder / which is easier?*
- What happens if I change ... ?
- If I know this, I also know ...

The prompts aim to elicit understanding of concepts, to get children thinking about how they can apply ideas to unfamiliar situations, and for teachers to assess whether children are making rich connections between different areas.

Italic bulleted text indicates questions the teacher could ask the children. →

- *What strategies could you use to add up these numbers?*

Other bulleted text indicates exemplar statements, questions or sentence stems from children. →

- I know that my rods will equal 20p because I have 1 more than 19p.

Where the first few words are **underlined**, this indicates the words the teacher could use to prompt children towards making the observation. →

- <u>I wonder</u> how many different ways there are to make a total of 19p.

Non-bulleted text indicates instructions to the teacher. →

Ask children to arrange their bead strings to show 19 beads.

A **yellow warning symbol** indicates important things to watch out for: either common misconceptions that may need addressing in case they lead children down a false path, or things that must be done before moving on to the next part of a task. →

⚠ Do children have a secure understanding of odd and even numbers? If not, ask them to create a list of odd and even numbers to check against.

The **Teacher Guide** tends to follow the same structure for each task. Below is an explanation of each section, using the task **Money Maze** for illustrative purposes.

Text in blue below indicates questions teachers may want to ask themselves when they get to each section of a task.

Introduction

This explains the aims of the task and how it delivers greater depth.

Money Maze

Year 1 Year 2

Introduction

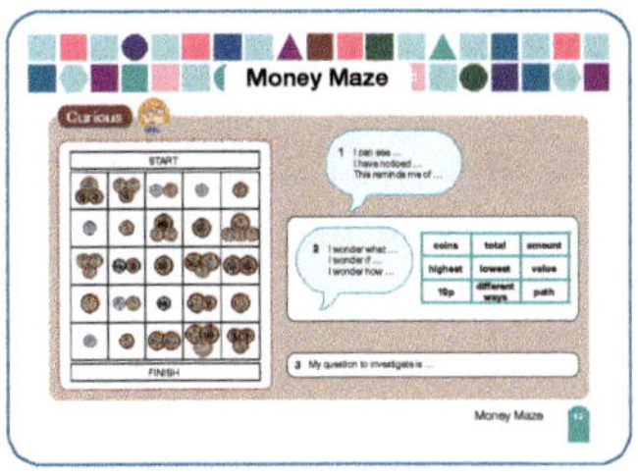

Money Maze provides children with the opportunity to rehearse and apply adding one-digit numbers within the engaging context of using money. Additional complexity is added through the use of multiple coins in each grid square and the possibility of a number of different lines of enquiry. Children may choose to calculate the value of each square before they start and assess whether this is an efficient way to work rather than adding all of the coins each time.

The task allows children to consider different resources to help them with their calculations, to keep track of their totals and to also provide proof. They use mental methods, jottings and concrete resources and consider in depth the suitability and efficiency of the strategies that they choose.

This task promotes greater-depth learning in a range of ways, supporting children to identify and articulate underlying mathematical structures. For example, in the Connect stage, greater depth is promoted by supporting children to choose an appropriate strategy and explain their choice, and understand that sometimes more than one answer is possible.

This **Teacher Guide** is supported by photocopiable **Pupil Resources** and **Teacher's Presentation** PowerPoint slides for whole-class or group display.

National Curriculum

All of the KS1 tasks can be done in either Year 1 or Year 2, but a particular Year group is suggested for each task.

Consider children's prior learning and experiences for each of the noted National Curriculum statements when deciding when to use each task.

Do children have sufficient skills and knowledge to access this task?

National Curriculum

Money Maze covers these Year 1 and Year 2 National Curriculum objectives.

Year 1 Addition and subtraction

- Add and subtract one-digit and two-digit numbers to 20, including zero.

Year 1 Measurement

- Recognise and know the value of different denominations of coins and notes.

Year 2 Addition and subtraction

- Add and subtract numbers using concrete objects, pictorial representations, and mentally, including: a two-digit number and ones; a two-digit number and tens; two two-digit numbers; adding three one-digit numbers.

Year 2 Measurement

- Find different combinations of coins that equal the same amounts of money.
- Solve simple problems in a practical context involving addition and subtraction of money of the same unit, including giving change.

The 6C cycle

The relative weightings of the 6C elements are given for each task to help teachers assess which tasks are suitable for their class at different times.

The 6C cycle

The grid on the right shows the relative weighting given to each of the 6 Cs in **Money Maze.**

Curious	☆☆
Connect	☆☆☆
Collaborate	☆☆
Communicate	☆
Create	☆
Consider	☆☆

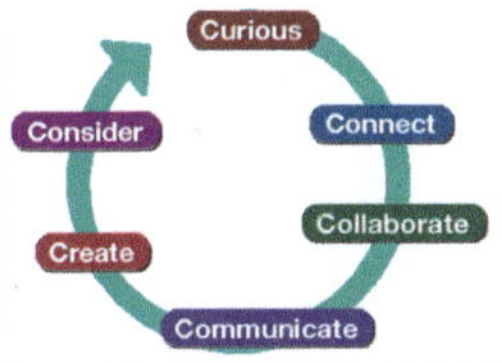

Problem-solving heuristics

For each task, specific problem-solving strategies are recommended for children to use during their investigation.

These strategies are then exemplified in the **Connect** section of the **Teacher Guide**.

Which strategies are children already aware of?

Is this an opportunity to teach a new strategy or apply a pre-existing skill?

Which strategy will be the focus of your modelling?

Problem-solving heuristics

The grid shows which problem-solving strategies children are likely to use when investigating the Connect tasks in **Money Maze.** These strategies are further exemplified in the Connect stage below (pages 54–57).

Draw a diagram	
Make a table / spot a pattern	
Make a model / act it out	✓
Try then improve	✓
Create a systematic list	✓
Solve a simpler but related problem	✓
Work backwards	✓

Key mathematical ideas

Notes are provided to support teachers' subject knowledge and outline the possible pedagogical pathways.

Are there any aspects you need to clarify before the teaching begins?

Key mathematical ideas

The aim of this task is to provide children with opportunities to rehearse and apply strategies for adding one-digit numbers. Children will get the most out of this task if they are secure in different methods for addition and subtraction.

Useful strategies could include:

- using near doubles: 5p + 6p = double 5p + 1p or double 6p – 1p
- looking for complements to 5: 3p + 4p + 2p could be rearranged to 3p + 2p + 4p
- looking for complements to 10: 3p + 5p + 7p could be rearranged to 3p + 7p + 5p
- regrouping to count on: 6p + 8p, regroup 6p into 2p and 4p, 8p + 2p + 4p
- regrouping to subtract: 15p – 8p, regroup 8p into 5p and 3p, 15p – 5p – 3p
- regrouping to subtract: 15p – 8p, regroup 15p into 10p and 5p, 10p – 8p + 5p
- counting on to subtract: 15p – 8p, counting on 2p from 8p to 10p and then counting on another 5p.

This task also gives children the opportunity to rehearse their knowledge of coins and how totals can be made with different combinations of coins.

Answers to some of the tasks are given on page 59, but other answers will vary depending on the lines of enquiry chosen.

Curious

For the Full 6C Pathway, advice is provided for teachers on how to introduce the question / problem to maximise curiosity.

Two progressive elements can be used:

- **Awe and wonder;**
- **Develop and deepen.**

Awe and wonder gets children talking about the question / problem and allows them to share their initial thoughts and start to become engaged.

'Stumble support' includes:

- examples of possible responses from children, with underlined words indicating how the teacher might prompt children towards these observations.

Awe and wonder

Slide 1

Display slide 1.

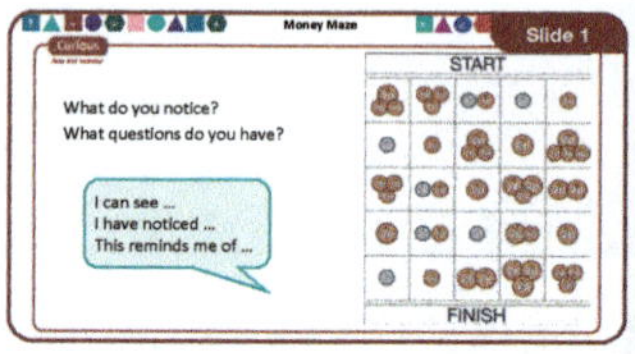

- I can see lots of coins.
- I can see a grid with 25 boxes.
- I have noticed that lots of the boxes have more than 1 coin in them.
- I have noticed that there are 1p, 2p and 5p coins.
- I have noticed that there is a start and finish.
- It reminds me of a treasure map a bit with the grid.

At this point, explain that you start from the top and choose a path to the finish line. Tell children that they can start on any square in the top row and finish on any square in the bottom row. They can move down, left, right and diagonally.

Curious

Develop and deepen aims to help children develop more mathematically accurate lines of enquiry, and to funnel children towards the pre-identified questions.

'Stumble support' includes:

- word banks to lift mathematical vocabulary and help direct children to appropriate lines of enquiry;
- examples of age-appropriate lines of enquiry.

How will the word bank be shared with children?

How will the possible lines of enquiry be collected? Whole class, small groups and so on. Do they need to be visible and shared with all to encourage children building on each other's ideas?

Will children all work on one line of enquiry or will they be able to select their preferred line of enquiry?

Develop and deepen

Encourage children to use the words in the word bank to develop and deepen their thoughts and questions.

coins	total	amount
highest	lowest	value
19p	different ways	path

Possible lines of enquiry linked to the word bank:

- I wonder what the total of each box is.
- I wonder what the highest / lowest value path to the finish line is.
- I wonder what the highest total I can make with 8 moves is.
- I wonder if two paths will have the same total.
- I wonder how many different ways there are to make a total of 19p.
- I wonder how many ways there are to make a different target.

Possible lines of enquiry that children may raise now, but should be held back until the **Create** task:

- I wonder what would happen if I changed the start and finish line and went across the board instead of down. What would the highest total be?

Consider whether / how to collate and display all the suggested questions, and whether / how to steer children towards selecting an appropriately challenging question.

At this point, confirm that you would like children to try to find paths that make a total of 19p.

Connect

The aim of the **Connect** task is to support children in using a range of problem-solving strategies, to try out different 'ways in' to the problem.

Guidance is given on the strategies best suited to the task and how they work in the given context.

Most **Connect** tasks will deliver a number of suitable alternative strategies. Children could work through all of the tasks suggested, or choose just one or two.

'Stumble support' includes:

- additional prompts, questions and suggestions for teachers to help children be successful with their chosen strategies.

Will children work through all of the Connect tasks, or just one or two?

What resources will need to be provided?

How will the 'stumble support' be made accessible?

Connect

Four tasks involving different strategies are suggested for children to make connections.

Both the main grid and the simpler grid are provided as printable slides (slides 6 and 7).

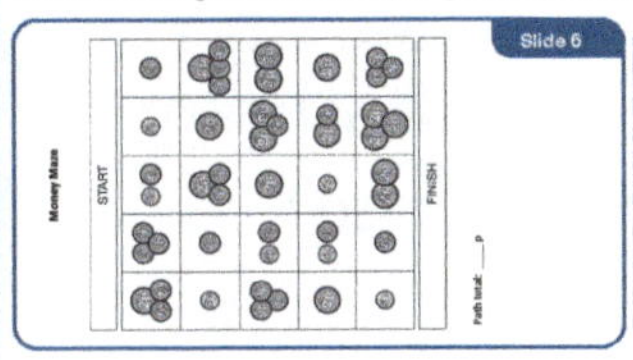

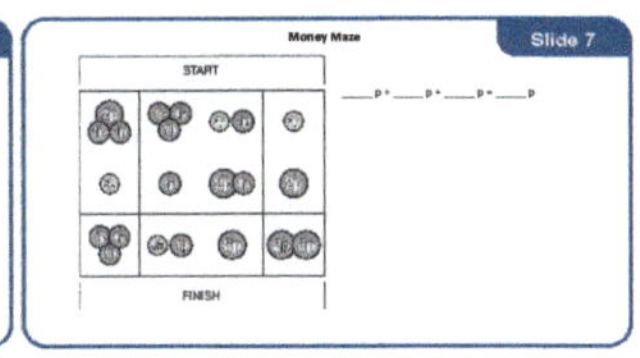

After printing, you may wish to laminate them for repeated use. Children can write the total in each box so that they do not have to work out the value each time they choose a different path.

Collaborate

The aim of the **Collaborate** task is for children to experience working with, or sharing results with, other children.

Various models of collaboration are suggested.

Which group-work skills will be a focus for your children?

Collaborate

Refer to the Collaborate task of the **Pupil Resource**.

Slide 14

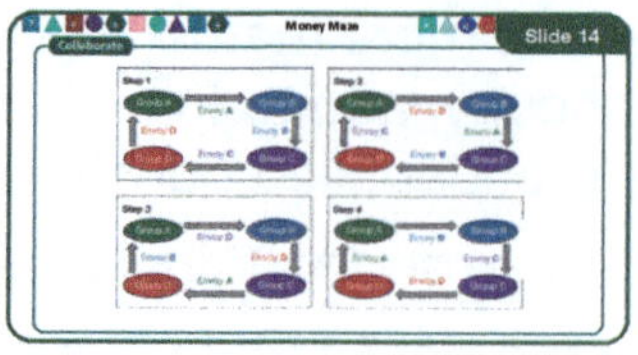

Display slide 14.

This task provides an opportunity for children to work independently, in pairs or in small groups to work through the maze.

As there are a number of solutions to this puzzle, this task provides a good opportunity for children to use the 'envoy' method. One child from each group is nominated as the envoy. Following discussion and exploration time, the envoy then moves to each of the other groups in turn, to ask questions about how they tackled the problem and the solutions they found, for example, the target total of 19p, highest value or lowest value path, before reporting back to their original group.

Communicate

The aim of the **Communicate** task is to get children to think about what they have discovered and how they are going to share their findings with others.

There may be opportunities for cross-curricular links to the National Curriculum.

'Stumble support' varies by task but may include:

- writing frames;
- word banks to deepen understanding;
- sentence stems.

How will respectful talk be modelled?

How and where will this be recorded?

Communicate

Refer to the Communicate task of the **Pupil Resource**.

Slide 15

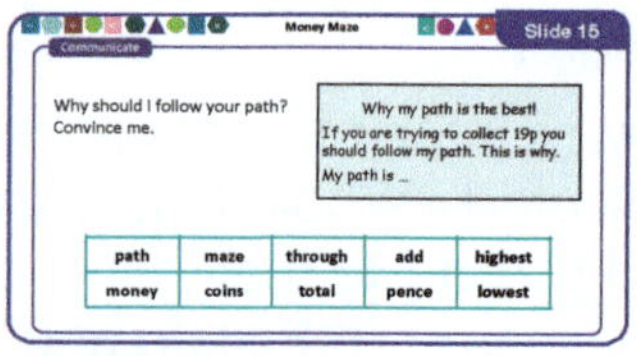

Display slide 15.

This task provides an opportunity for the children to create a short piece of persuasive writing. Alternatively, children could give a short speech about why their path is the best. They could use sound buttons or audio devices to record this.

Encourage them to use the vocabulary from the word bank.

path	maze	through	add	highest
money	coins	total	pence	lowest

Create

The aim of the **Create** task is to inspire children to continue with developing their own questions to further deepen learning.

Examples of further lines of enquiry are provided in the **Teacher Guide** for each task.

Unanswered questions from the **Curious** stage can be returned to and engaged with.

'Stumble support' includes:

- sentence stems
- word banks to extend vocabulary.

Were any additional lines of enquiry suggested in the Curious task that could be pursued now?

Create

Refer to the Create task of the **Pupil Resource**.

- *Were there any questions asked at the start of the task that you now want to think about?*
- *Is there anything new that you are wondering now?*

Children can use the word bank to prompt further questions of their own.

across the board	change	start
finish	fewest moves	highest total

- I wonder what would happen if I changed the start and finish and went across the board instead of down. What would the highest total be?

Consider

The aim of the **Consider** task is to get children to reflect on what they have learned.

- Cognitive domain – what have they learned?
- Affective domain – how do they feel about the learning?
- Metacognitive domain - how did they learn it?

'Stumble support' includes:

- additional sentence stems.

Which of the suggested sentence stems will you encourage children to consider and complete?

Refer to the Consider task of the **Pupil Resource**.

Encourage children to think about the strategies that they have used. The prompts below may help get children started.

Cognitive learning domain

- The trickiest part of solving my question was …
- The new strategies I learned were …

Affective domain

- The fun bit of my learning was … It was fun because …
- It was easy / hard to keep track of the different paths because …
- I am / I am not pleased with my work because …

Metacognitive domain

- I used … to help me. It was helpful because …
- Next time, I will …
- A top tip I would give for working well as a group is …

Consider

Answers

Where appropriate, worked solutions and answers are provided.

For some tasks, these will be presented on the final page of the **Teacher Guide** for that task. For others, answers are given in situ throughout, because of the cumulative nature of the tasks.

Answers

There are many possible paths to 19p. For example:

4p + 5p + 3p + 2p + 5p = 19p

6p + 2p + 5p + 3p + 3p = 19p

1p + 5p + 4p + 2p + 3p + 4p = 19p

Solve a simpler but related problem

Four possible paths to 11p:

4p + 1p + 6p = 11p

6p + 3p + 2p = 11p

3p + 5p + 3p = 11p

5p + 2p + 4p = 11p

How to use the Pupil Resources

Full Pathway Curious page

Curious question / problem
This acts as an *aide memoire* for the rest of the cycle.

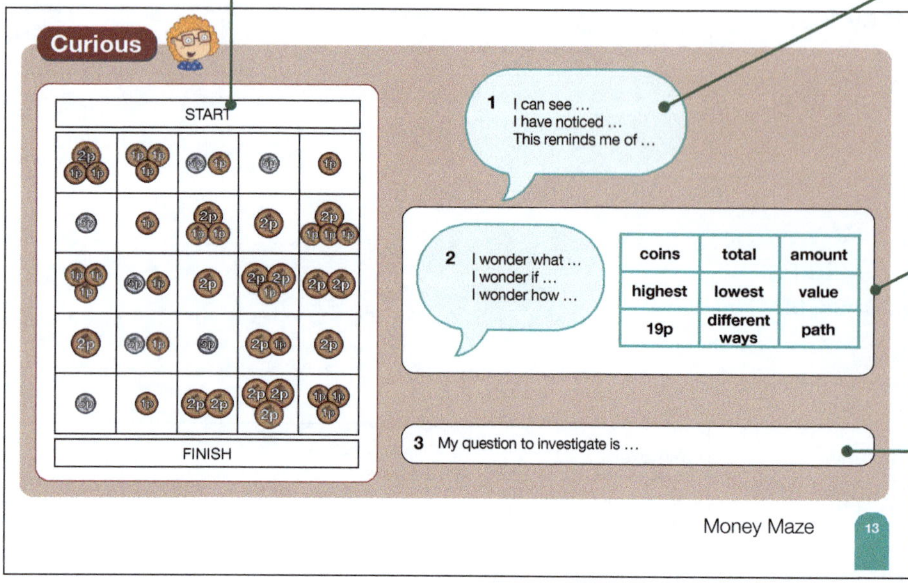

Awe and wonder
Sentence stems to support children to articulate their initial observations and ideas.

Develop and deepen
A word bank to boost mathematical vocabulary and support mathematically accurate questions.
Sentence stems to support children to articulate their questions.

Chosen line of enquiry
A prompt for children to select a line of enquiry they would like to pursue.

Express Pathway Curious page

(See note on page 9 about tasks that are allocated to Year 1 only.)

Curious question / problem
This acts as an *aide memoire* for the rest of the cycle.
The question / problem is presented in a more focused way.

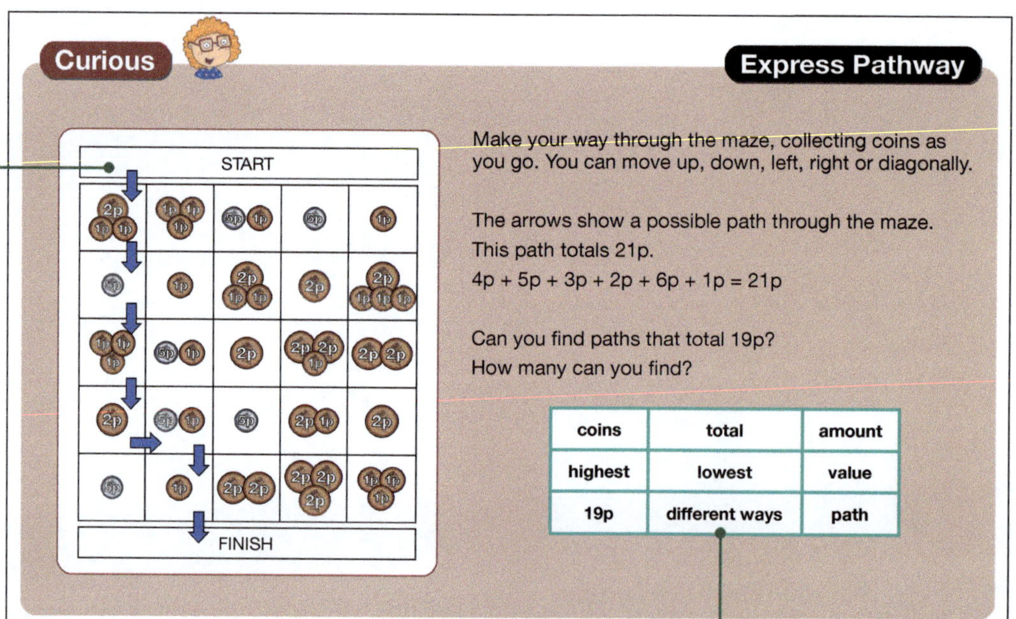

A word bank to boost mathematical vocabulary.

The remaining 5 Cs, common to both pathways

Connect
Children are presented with a range of strategies. They could select their preferred option, or try out more than one.

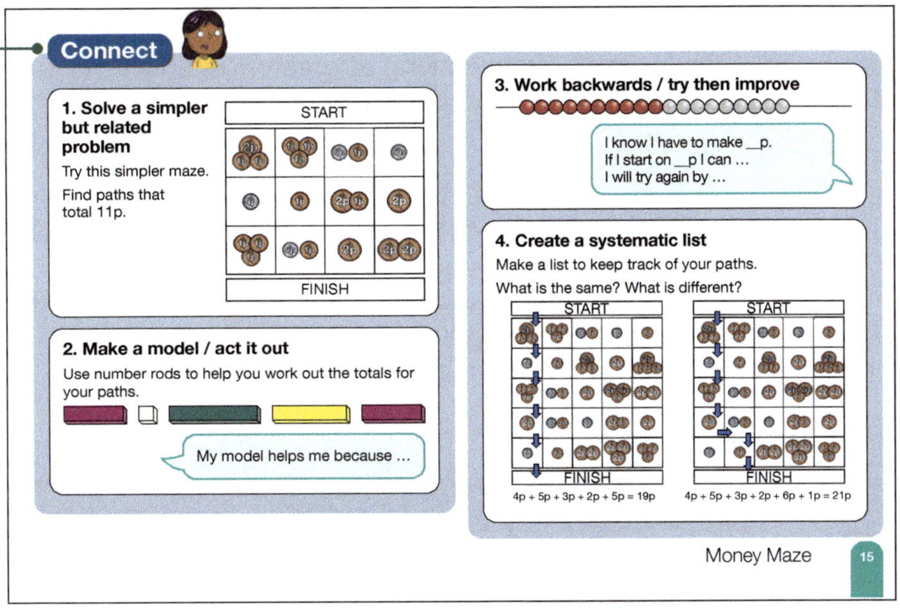

Collaborate
Children are prompted to work with others, and consider helpful ways of working.

Create
Children are prompted to continue developing their own questions to further deepen their learning. A word bank encourages mathematically accurate language.

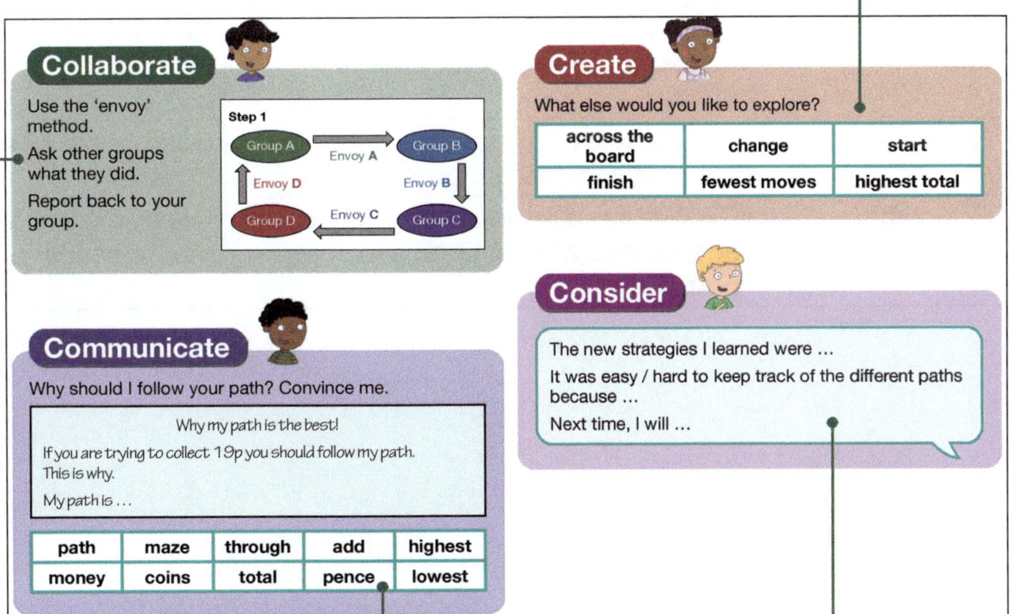

Communicate
Children are asked to explain their findings clearly, and are given suggestions for how their ideas could be shared. Word banks, sentence stems and writing frames are provided where appropriate.

Consider
Sentence stems guide children in reflecting on what they have learned, how they learned it and how they feel about their learning.

Assessment rubric

The assessment rubric has been designed to identify the stepping stones of attainment across the 6Cs. For a child to be described as working at greater depth, they need to be performing highly in all components. Most children working at greater depth will have a mixed profile between consolidating and extending. The rubric should be used with the whole of the class.

Using a rubric can help with:

- developing self-regulation: children can use it to help plan or monitor their own learning;
- direct teaching: it moves the focus from task completion to learning skills, knowledge and behaviour;
- providing a learning pathway: when used over time, it can help to identify gaps and next steps so feedback can be fine-tuned and highlighted as a teaching opportunity; this feedback and teaching then loops back into the next task.

Key Stage 1 assessment rubric

	Curious	
	Awe and wonder	*Develop and deepen*
Not evident		
Entering	Makes limited contributions. Contributions made are not connected to the task.	Makes unconnected statements / questions when prompted.
Beginning	Makes simple contributions when prompted, e.g. recalls key fact or simple pattern.	Contributes statements rather than questions with limited mathematical content, sometimes needs prompting.
Developing	Contributes to discussion, e.g. notices a common property or pattern.	Contributes statements rather than questions, with some mathematical content.
Consolidating	Contributes to discussion, identifies connections and patterns.	Offers a single mathematical question.
Extending	Contributes to discussion. Notices and queries connections and patterns.	Offers more than one good mathematical question.

	Connect		Collaborate		
	Application	*Reasoning*	*Individual responsibility*	*Teamwork*	*Respect for others*
Not evident					
Entering	Has trouble figuring out how to solve the problem. Needs heuristic to be chosen for them.	Not sure if their answer is correct or not. Describes what they did.	Often or always relies on others to do the work.	Only shares ideas and suggestions when prompted to.	Impolite or unkind to the group.
Beginning	Needs some help to choose an appropriate heuristic.	With help can figure out if they have a correct answer. Attempts an explanation, but it is incorrect.	Participates but needs to be prompted and encouraged throughout.	Needs occasional prompting to share ideas and suggestions.	Sometimes ignores the feelings of others, e.g. interrupting or ignoring ideas.
Developing	Able to select an appropriate heuristic from a predefined list.	Provides a correct answer but may get confused on the way. Offers an explanation but it is incomplete / unsystematic.	Participates but needs occasional prompting. Gets upset if they do not get to do what they want.	Shares ideas and makes suggestions but gets upset if they are rejected by the group.	Usually polite and kind to the group.
Consolidating	Selects an appropriate heuristic. Can explain why they have chosen it drawing on previous experience.	Provides a correct answer. Explains the pattern or rule and how it works. A systematic approach is present, if relevant.	Participates within the group without arguing.	Offers suggestions and ideas.	Polite and kind to all group members.
Extending	Selects the most appropriate heuristic and can explain choice.	Understands that sometimes more than one answer is possible.	Participates readily and is willing to change roles.	Offers ideas and suggestions. Accepts if they are rejected by the group.	Polite, kind and encouraging to group members.

	Communicate		Create
	Clarity	*Mathematical language, images and diagrams*	
Not evident			
Entering	Communication is confusing and difficult to follow.	No mathematical language.	Not able to suggest any further questions.
Beginning	Communication is hard to follow, but can answer questions about learning if prompted.	Attempts to use some mathematical language.	Needs help to develop new relevant questions.
Developing	Communication is hard to follow in places.	Uses some mathematical language.	Builds on others' questions, changing them slightly.
Consolidating	Communication is clear and easy to follow.	Mathematical language is accurate.	Creates their own questions.
Extending	Communication is clear, accurate and well organised.	Mathematical language is normally accurate supported with a diagram / image / model.	Creates their own questions that are creative and inventive.

	Consider		
	Cognitive	*Affective*	*Metacognition*
Not evident			
Entering	Confuses learning outcomes with title of the task.	Only states that the learning was fun or not, struggles to explain why.	Struggles to identify anything they did when they got stuck. Passively waits for an adult to support them.
Beginning	Attempts to describe what they did, but it is unclear.	Can state that the learning was fun or not; with prompting they can offer a simple reason why.	Requires prompting to identify which learning behaviours they found helpful when they got stuck and if anyone else helped them.
Developing	Able to describe what they did.	Able to identify when things got challenging. Identifies the learning as fun or not, and can explain why.	Identifies which learning behaviours they found helpful when they got stuck and, if anyone else helped them, how.
Consolidating	Able to identify what they have learned.	Able to explain how their feelings changed throughout the task. Describes the learning as engaging or not engaging.	Explains which learning behaviours were useful and which were not.
Extending	Able to identify more than one learning outcome.	Reflects on how they were feeling throughout the task and why. Able to identify what was and was not engaging about the task.	Explains which learning behaviours were useful and which were not. Includes some ideas on what they could do differently next time.

I Can Read Your Mind

Year 1

Introduction

I Can Read Your Mind provides an exciting opportunity for children to explore the power of number patterns. In order to get the most out of the task, children should be able to subitise (instantly recognise the number of dots on the dice face without having to count).

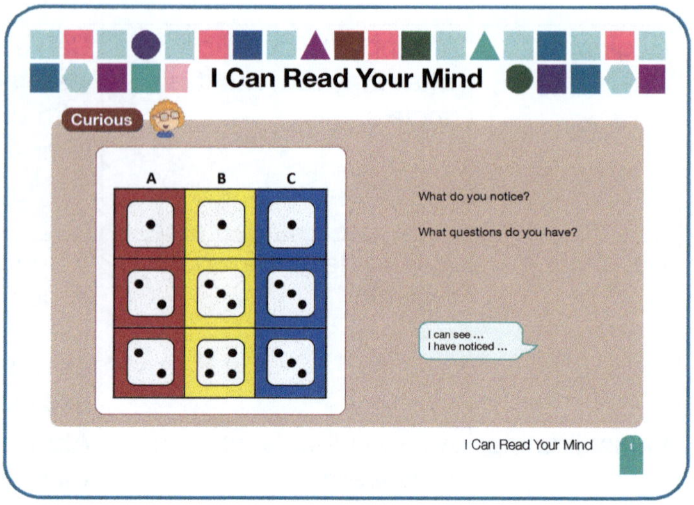

During the task children will observe the structure of the grid and the similarities and differences of the positions of each number to discover how the 'mind-reading trick' works, and then apply and extend this knowledge to become mathematical mind readers themselves.

This task promotes greater-depth learning in a range of ways, supporting children to identify and articulate underlying mathematical structures. For example, in the **Communicate** stage, greater depth is promoted by supporting children to communicate their thoughts clearly, accurately and in a well-organised way.

This **Teacher Guide** is supported by photocopiable **Pupil Resources** and **Teacher's Presentation** PowerPoint slides for whole-class or group display.

National Curriculum

I Can Read Your Mind covers these Year 1 National Curriculum objectives.

Number and place value

- Read and write numbers from 1 to 20 in numerals and words.

The 6C cycle

The grid on the right shows the relative weighting given to each of the 6 Cs in **I Can Read Your Mind**.

Curious	★★
Connect	★★★
Collaborate	★★
Communicate	★★★
Create	★★
Consider	★★

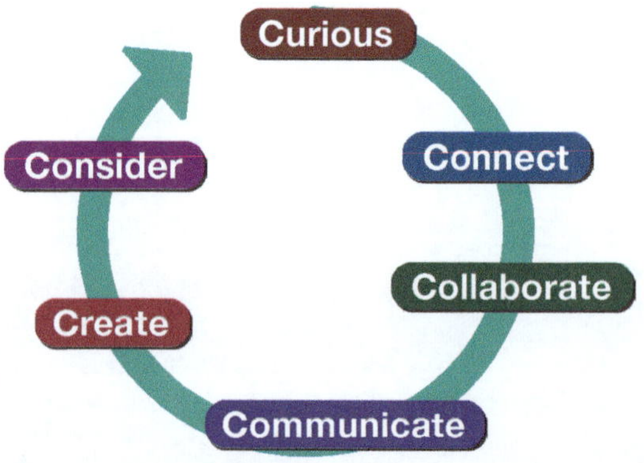

Problem-solving heuristics

The grid shows which problem-solving strategies children are likely to use when investigating the **Connect** tasks in **I Can Read Your Mind**. These strategies are further exemplified in the **Connect** stage below (pages 28–30).

Draw a diagram	
Make a table / spot a pattern	✓
Make a model / act it out	✓
Try then improve	
Create a systematic list	
Solve a simpler but related problem	✓
Work backwards	

Key mathematical ideas

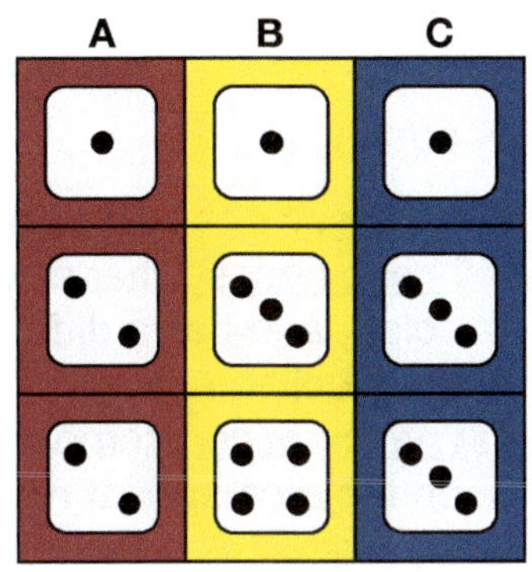

The crucial element to this 'trick' is the arrangement of the dice across the columns. It is vital that each dice has a unique placement so that answers to the questions you pose indicate which number is being thought of. As long as no other dice is placed in exactly the same combination of column(s), you will be right. For example:

1 is the only number in all three columns.

2 is the only number to just appear in the red column A.

3 is the only number to just appear in both the yellow B and blue C columns.

4 is the only number to just appear in the yellow B column.

- *If I answer 'Yes' to your question 'Is your number in column B?', but 'No' to 'Is your number in column A?' and 'No' to 'Is your number in column C?', you know that I am thinking of number 4.*

Therefore, when children are creating their own 'mind reading' grids, it is important that they work strategically and keep track, for example, with a table, of the combinations to ensure each number has a unique placement. For example:

A	B	C	D
	1		1
2		2	2
3			
		4	4
5	5	5	5
		6	

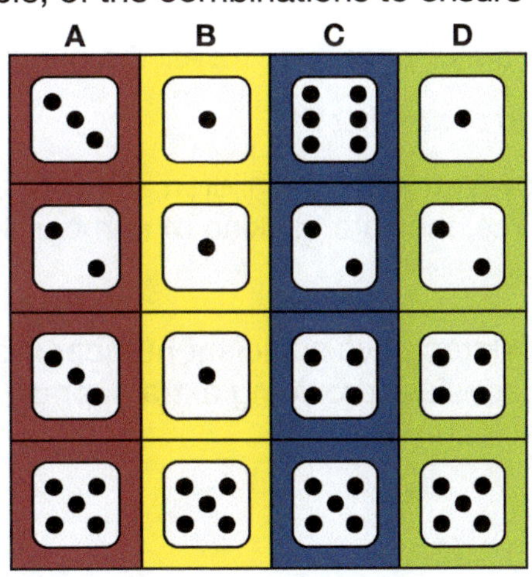

It is important for this task that children can subitise, or instantly see how many. If children do not have sufficient number recognition skills to identify and record numerals, you may wish to build up this fluency by playing games with dot pattern cards, such as memory, dominoes or snap.

The key to the 'tricks' are given in situ throughout the tasks.

Teaching guidance

Teaching guidance is given below.

The **Teacher's Presentation** PowerPoint file has slides specifically mentioned in this teaching guidance. Some of the slides include notes (within the PowerPoint file) providing 'stumble support' suggestions, that may help you to support children in their problem-solving thinking and enable access for children who are less confident.

Curious

Awe and wonder

Slide 1

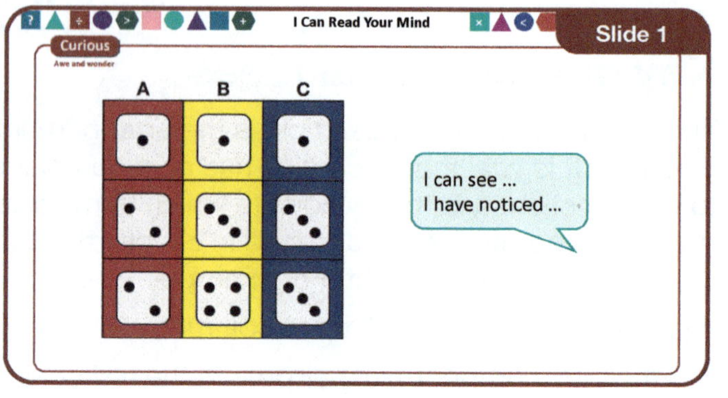

Display slide 1.

Ask children to share what they are noticing or wondering.

- I can see lots of dice on a coloured grid.
- I have noticed that each column is a different colour and has a different letter. The columns are labelled A, B, C.
- I have noticed some of the numbers shown on the dice appear more than once.
- I can see 1 is on the board three times.
- I can see that the highest number on the dice is a 4.

Now perform the 'mind reading' trick with one of the children.

- *Choose a number between 1 and 4.* [For example, the child chooses 2.]
- *Is your number in red column A?* [Yes.]
- *Is your number in yellow column B?* [No.]
- *Is your number in blue column C?* [No.]

Pretend to think carefully and then reveal the number based on the answers to your questions.

- *You are thinking of number 2!*

Do not explain how you know. If children answer Yes, Yes, Yes, they are thinking of number 1, if they answer No, Yes, Yes, they are thinking of number 3, if they answer No, Yes, No, they are thinking of number 4.

⚠ Make sure children have sufficient number recognition skills to identify and record numerals. If they do not, consider providing extra support such as a number mat that shows a dot pattern alongside each numeral.

Develop and deepen

Encourage children to use the words in the word bank to develop and deepen their thoughts and questions.

same	**dice**	**number**
both	**only**	**different**
colours	**column**	**all**

Possible lines of enquiry:

- I wonder how the correct dice / number is found every time.
- I wonder why some dice / numbers are on the board more than once.

Possible lines of enquiry linked to the word bank:

- I wonder if I can use the columns to work out the trick.
- I wonder what is the same and what is different about each number in the grid.
- I wonder if any of the numbers are only in one column.

Possible lines of enquiry that children may raise now, but should be held back until the **Create** task:

- I wonder if I could create my own mind reading trick with larger numbers.
- I wonder if the trick works with other things (for example, shapes, colours, words).

Consider whether / how to collate and display all the suggested questions, and whether / how to steer children towards selecting an appropriately challenging question.

Pupil Resources

Distribute the **Pupil Resources** for **I Can Read Your Mind** to the children to help support them in their exploration of the different tasks.

To provide additional flexibility in the classroom, each of the **Pupil Resources** is also provided in the **Teacher's Presentation** PowerPoint file (see slides 2–5), so you can display these at any time if you wish.

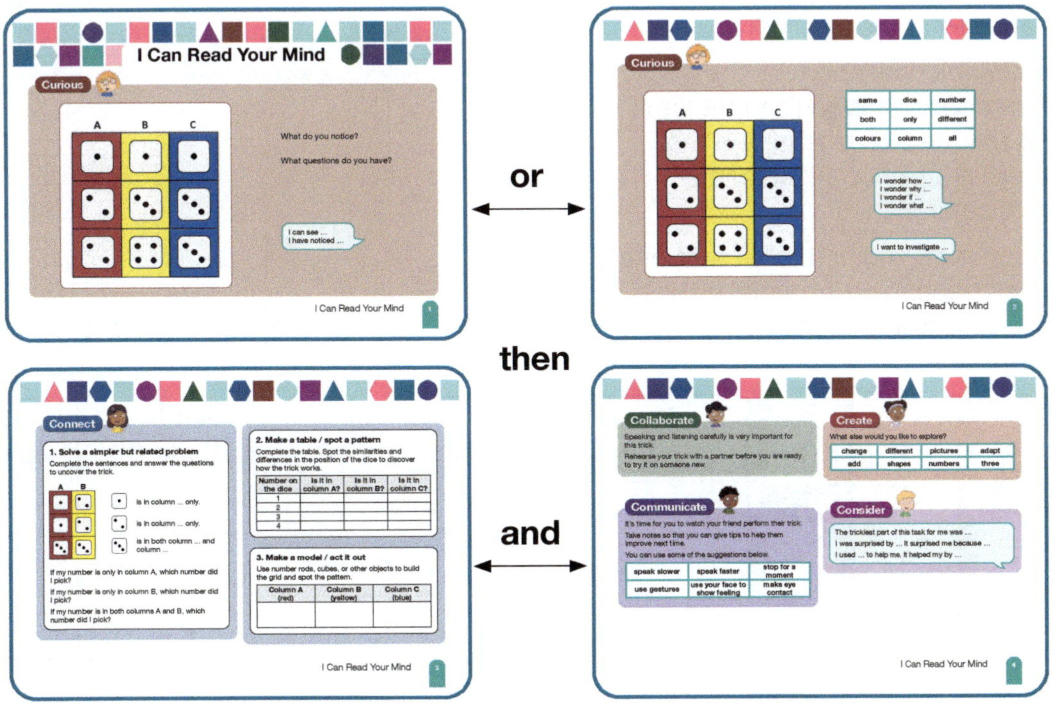

Three tasks involving different strategies are suggested for children to make connections.

1. Solve a simpler but related problem

Slide 6

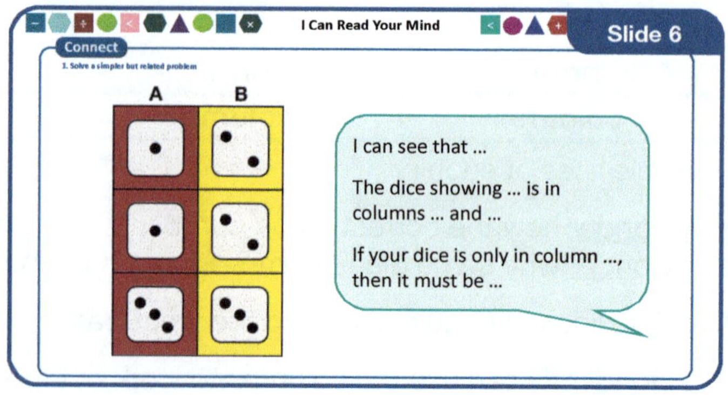

Display slide 6. This shows a 2 by 3 grid featuring 3 different dice rather than 4. Making a table or a model in order to help spot a pattern, as detailed below, can also be used as strategies to discover the 'trick' for this simpler problem as well for the full 4 by 3 grid.

Encourage children to use the sentence frames to support their thinking and to focus on one dice at a time if necessary.

- *Where is there a dice showing the number 1?*
- I can see that the 1 dice is only in the red A column.
- *What about a dice showing the number 3?*
- The dice showing 3 is in the red A column and the yellow B column.
- *If your dice is only in the yellow B column, what number must it be?*

1 is the only number to just appear in the red A column. 2 is the only number to just appear in the yellow B column. 3 is the only number which is in both the red A and yellow B columns.

If the answer is 'Yes' to 'Is your number in column A?' and 'No' to 'Is your number in column B?', the person is thinking of number 1.

If they answer No, Yes, they are thinking of number 2.

If they answer Yes, Yes, they are thinking of number 3.

2. Make a table / spot a pattern

Slide 7

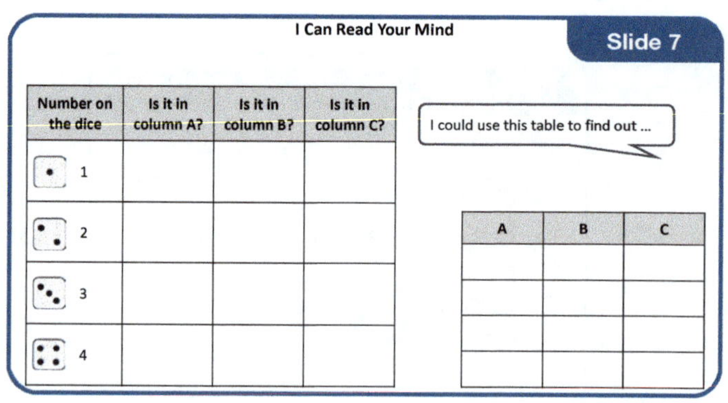

Print slide 7 and distribute to children.

To be able to unlock how and why the 'trick' works, children need to recognise that each number has a unique placement, and that no two numbers are placed in the same combination of columns.

Ask them to complete the first table with either a tick or a cross depending on whether the number on the dice is in the column.

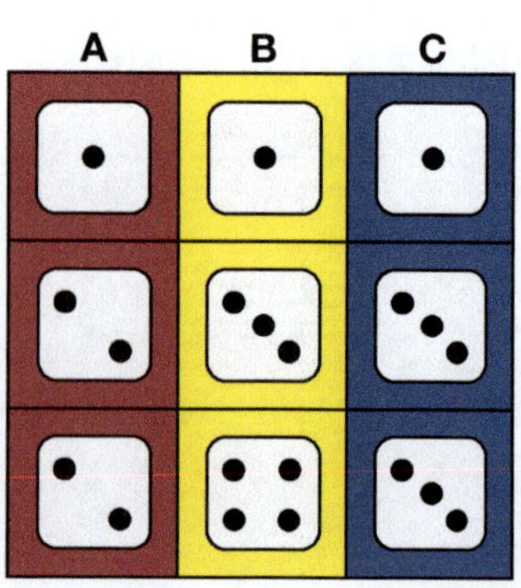

Number on the dice	Is it in column A?	Is it in column B?	Is it in column C?
1	✓	✓	✓
2	✓	✗	✗
3	✗	✓	✓
4	✗	✓	✗

Some children may need help to unlock the structure of the grid by making the placement of each number clearer. Ask them to fill in the second table with the position of each number.

A	B	C
1	1	1
2		
	3	3
	4	

Slide 8

Display slide 8 which shows the completed tables. Some children may need direct questions to draw attention to the structure.

- *Which number is the only one to be in just the red column A?*
- There's only one tick for the number 2 and that's under the red column A. So 2 is the only number that is just in column A.
- *Which number occurs in all columns?*
- There are three ticks for number 1. So number 1 is in all three columns.
- *Are there any numbers which are in exactly the same columns?*

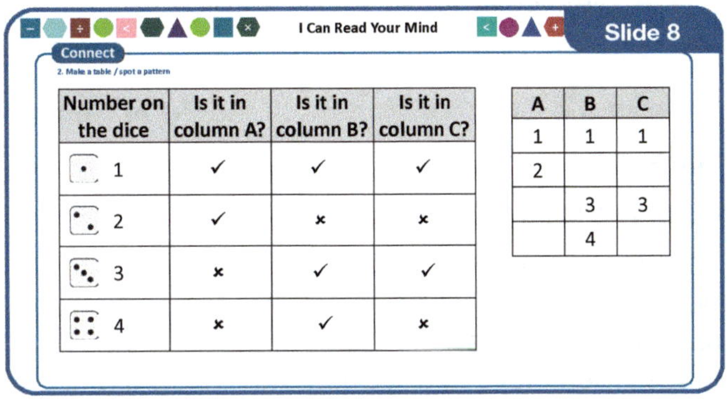

Number on the dice	Is it in column A?	Is it in column B?	Is it in column C?
1	✓	✓	✓
2	✓	×	×
3	×	✓	✓
4	×	✓	×

A	B	C
1	1	1
2		
	3	3
	4	

Ask children to think about the questions they can now ask and how they can use the answers to help work out which number has been chosen.

- You said it is only in the yellow column B. That means you must have chosen a 4!
- It must be a 1 because it is in all of the columns!
- If your number is only in the yellow B and blue C columns then it must be a 3!

3. Make a model / act it out

Additional resources required: concrete resources to represent numbers (for example, number rods, interlocking cubes or Numicon Shapes)

Slide 9

Print out slide 9 and distribute to children.

Explain that building a model can also lead them to spot the pattern and recognise the unique positioning of each number. Ask them to use concrete resources to build the grid and establish the patterns.

Some children may initially benefit from focusing on fewer numbers by using the 'Solve a simpler but related problem' **Connect** task first to allow them to focus on the structure of the grid.

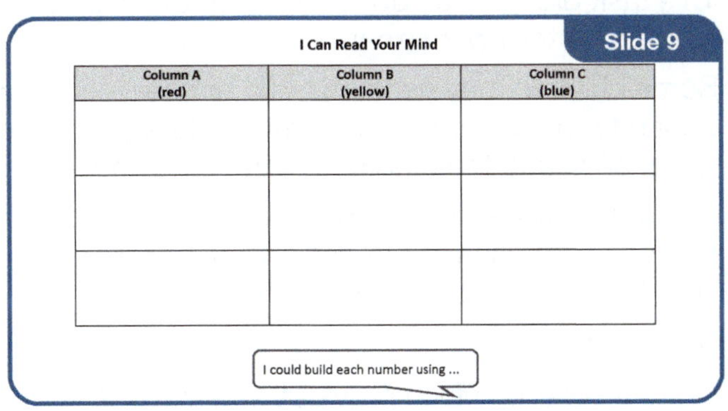

Connect

Slide 10

Display slide 10 which shows an example grid completed with number rods.

Some children may need questions to draw attention to the structure. Use the example with number rods on the slide and ask questions such as:

- *What is the same and what is different about the arrangement of each number?*
- *Are there any numbers which are in exactly the same columns?*

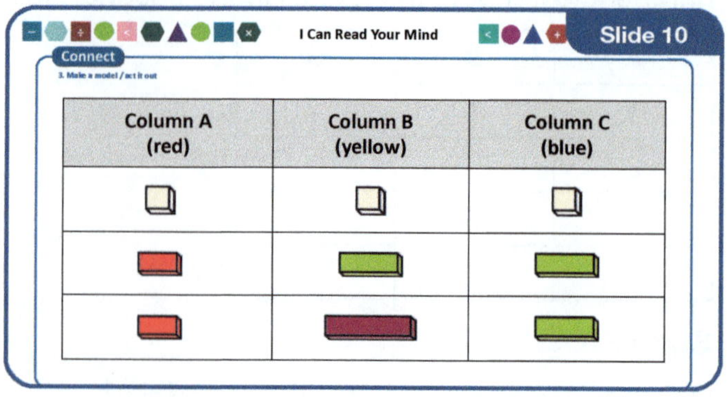

Collaborate

Refer to the **Collaborate** task of the **Pupil Resource**.

Slide 11

Display slide 11.

This task provides an opportunity for children to work in pairs to try out and perform the 'mind-reading trick'. This includes both at the initial stage when they are exploring the structure of the trick to see how it works and also if the children go onto create their own 'mind-reading' grids.

Ask children to compare the way in which they performed the 'trick'.

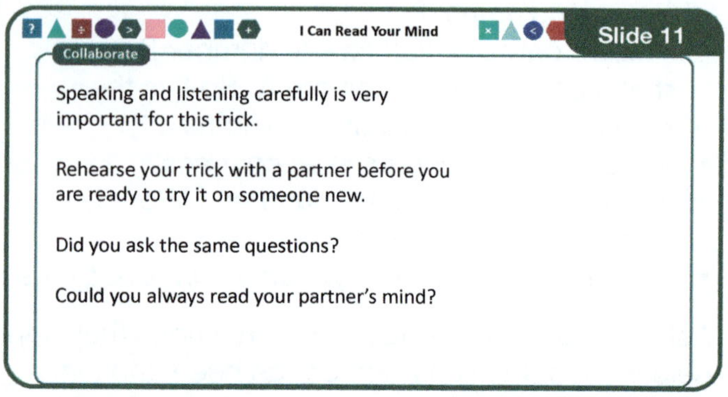

- *Did you ask the same questions?*
- *Was your 'mind-reading' successful each time?*

Children could then work in groups to give constructive feedback to improve their presentations as detailed in the **Communicate** task.

Communicate

Refer to the **Communicate** task of the **Pupil Resource**.

The 'mind reading trick' relies on excellent speaking and listening as the answers to the questions are needed to be able to identify the chosen number successfully.

This task provides an opportunity for the children to perform their tricks and gather feedback in order to improve their performance.

Some children may need an adult or peer to repeatedly model the trick to gain a stronger understanding of how to communicate well, both in terms of speaking (for example, asking the questions), and listening (for example, gathering the information required from the answers).

Children may need support to keep a record of the responses during the 'trick' to give them time to identify the chosen number.

Slide 12

Display slide 12 which provides suggestions they can use to give feedback to their peers on what went well and what could be improved.

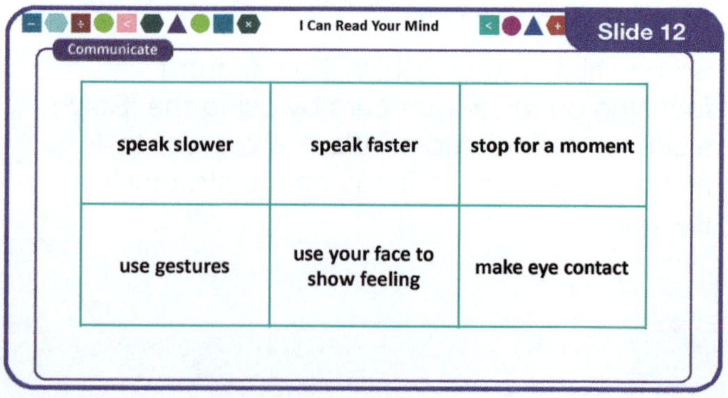

Refer to the **Create** task of the **Pupil Resource**.

Slide 13

Display slide 13.

- *Were there any questions asked at the start of the task that you now want to think about?*
- *Is there anything new that you are wondering now?*

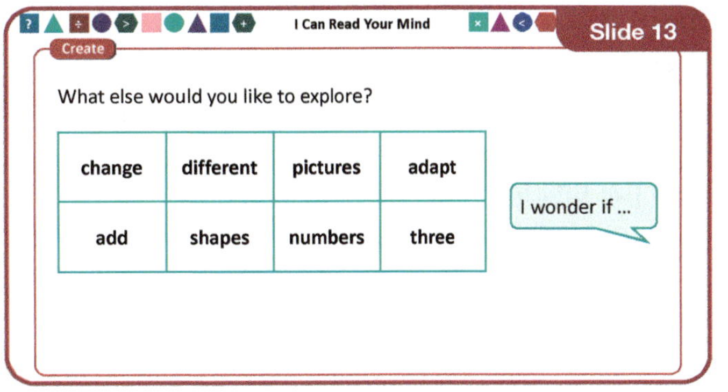

Children can use the word bank to prompt further questions of their own.

change	**different**	**pictures**	**adapt**
add	**shapes**	**numbers**	**three**

- I wonder if I could change the number of columns.
- I wonder if the trick works with other things (for example, shapes, colours, words).
- I wonder if I could try a different type of grid.
- I wonder if I could create my own mind reading trick with larger numbers.

Make sure children understand that the principle of each number / shape / picture having to have a unique positioning must still apply when creating tricks with something other than dice. For example:

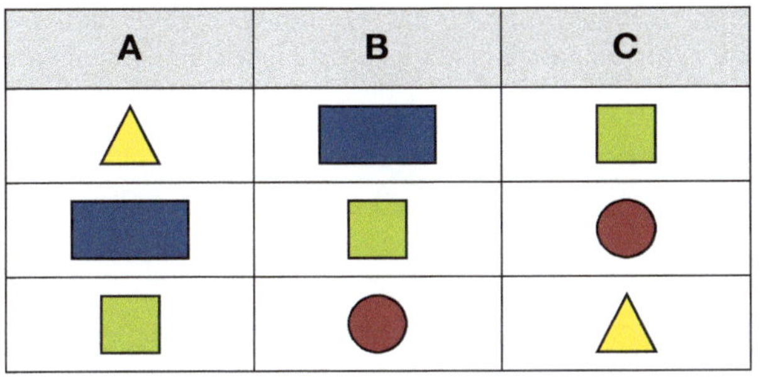

Consider

Refer to the **Consider** task of the **Pupil Resource**.

Slide 14

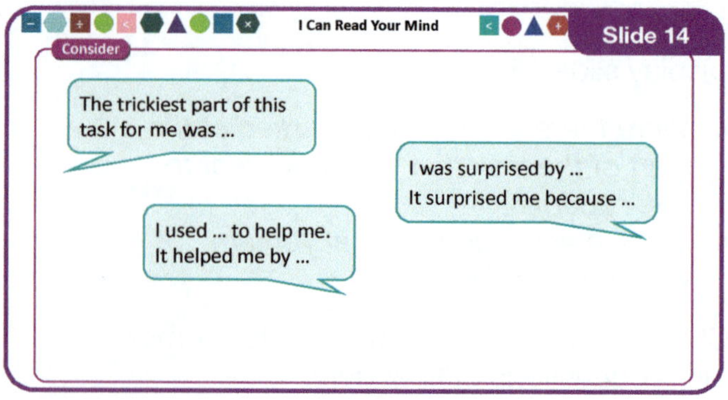

Display slide 14.

Encourage children to think about the strategies that they have used. The prompts below may help get children started.

Cognitive learning domain

- The trickiest part of this task for me was …
- I learned …

Affective domain

- The fun bit of my learning was … It was fun because …
- I said 'aha' when … I said it because …
- I was surprised that … because …

Metacognitive domain

- I got stuck when … I unstuck myself by …
- I used … to help me. It helped me by …
- Next time, I will …

Hidden Dragon Eggs

Year 1

Introduction

Hidden Dragon Eggs provides children with an opportunity to apply their developing understanding of odd and even numbers. Children focus on the structure of numbers between 1 and 16 by experimenting with the different ways the numbers can be represented on a 4 by 4 grid.

The open-ended nature of the task means children will benefit from working systematically. They can create a model to investigate the structure of odd and even numbers, or start with a smaller grid (for example, 2 by 3), before applying their reasoning to the full problem. The **Communicate** game allows children to rehearse and refine their use of positional language.

This task promotes greater-depth learning in a range of ways, supporting children to identify and articulate underlying mathematical structures. For example, in the **Collaborate** task, greater depth is promoted by supporting children to be polite and encouraging to group members, and to offer their own ideas and accept it if they are rejected by the group.

This **Teacher Guide** is supported by photocopiable **Pupil Resources** and **Teacher's Presentation** PowerPoint slides for whole-class or group display.

National Curriculum

Hidden Dragon Eggs covers these Year 1 National Curriculum objectives.

Number and place value

- Identify and represent numbers using objects and pictorial representations including the number line, and use the language of: equal to, more than, less than (fewer), most, least.
- *Pupils practise counting as reciting numbers and counting as enumerating objects, and counting in twos, fives and tens from different multiples to develop their recognition of patterns in the number system (for example, odd and even numbers), including varied and frequent practice through increasingly complex questions.* [non-statutory guidance]

The 6C cycle

The grid on the right shows the relative weighting given to each of the 6 Cs in **Hidden Dragon Eggs**.

Curious	★★
Connect	★★
Collaborate	★★★
Communicate	★★
Create	★★
Consider	★★

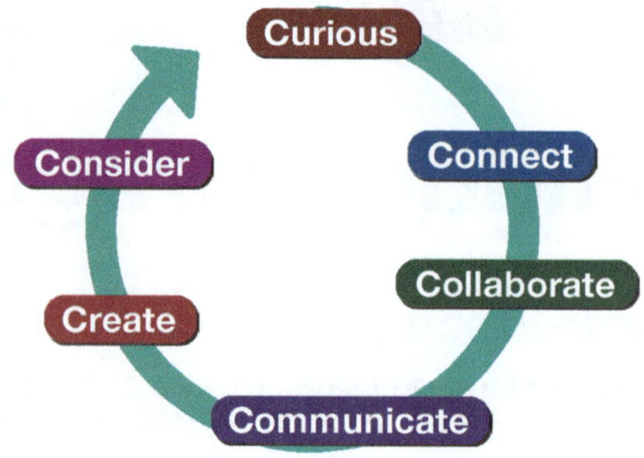

Problem-solving heuristics

The grid shows which problem-solving strategies children are likely to use when investigating the Connect tasks in **Hidden Dragon Eggs**. These strategies are further exemplified in the Connect stage below (pages 37–39).

Draw a diagram	
Make a table / spot a pattern	✓
Make a model / act it out	✓
Try then improve	
Create a systematic list	✓
Solve a simpler but related problem	✓
Work backwards	

Key mathematical ideas

Before exploring this task, it is essential that children are secure with the concept and recall of odd and even numbers.

Although It can be argued that zero is an even number (it is an integer multiple of 2, and when we divide it by 2, it results in a whole number: $0 \div 2 = 0$), children do not have to consider zero as an even number for the purpose of this task.

In the task, each row and column must have an even number of eggs or no eggs at all.

It is not possible to place an odd number, because at least one column and / or row would always have an odd number. For example, here are three different ways 5 eggs could be arranged. In each diagram, at least one column has only 1 egg.

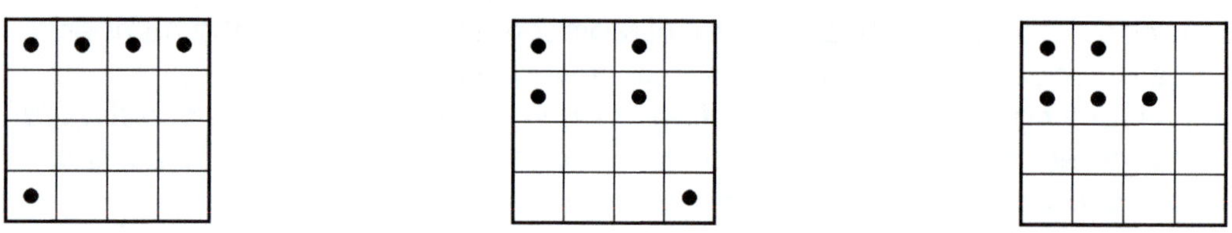

Some children may think at first that it is possible with all even numbers. However, with 2 eggs it is not possible, as at least two rows or columns would have only one egg.

•	•		

		•	
		•	

	•		
			•

There are also no possible solutions for 6, 10 or 14 eggs, as at least two rows or columns will be left with an odd number.

6 eggs

•		•	
	•		•
•		•	

10 eggs

•	•	•	•
•	•	•	•
•			•

14 eggs

•	•	•	•
•	•	•	•
•	•	•	•
		•	•

There are numerous possible solutions for 4, 8 and 12 eggs.

4 eggs

•	•		
•	•		

There are either 2 eggs or no eggs in each row and column.

8 eggs

•	•	•	•
•	•	•	•

There are either 2 eggs, 4 eggs or no eggs in each row and column.

12 eggs

•	•	•	•
•	•	•	•
		•	•
		•	•

There are either 2 eggs or 4 eggs in each row and column.

It is also possible for 16 eggs when all the boxes are filled as there are 4 eggs in each column and each row.

Possible answers to the tasks are given on page 41.

Teaching guidance

Teaching guidance is given below.

The **Teacher's Presentation** PowerPoint file has slides specifically mentioned in this teaching guidance. Some of the slides include notes (within the PowerPoint file) providing 'stumble support' suggestions, that may help you to support children in their problem-solving thinking and enable access for children who are less confident.

Curious

Awe and wonder

Slide 1

Display slide 1.

Ask children to share what they are noticing or wondering.

- I can see there are 16 squares.
- I have noticed it is 4 + 4 + 4 + 4.
- This reminds me of a treasure map.
- I have noticed it doesn't say how many eggs there are.
- I can see that there can be an even number of eggs, so there could be 4 or 2.
- I have noticed there can be no eggs in a row or column, that's zero.

⚠ Do children have a secure understanding of odd and even numbers? If not, ask them to create a list of odd and even numbers to check against.

Develop and deepen

Slide 2

Display slide 2.

Encourage children to use the words in the word bank to develop and deepen their thoughts and questions.

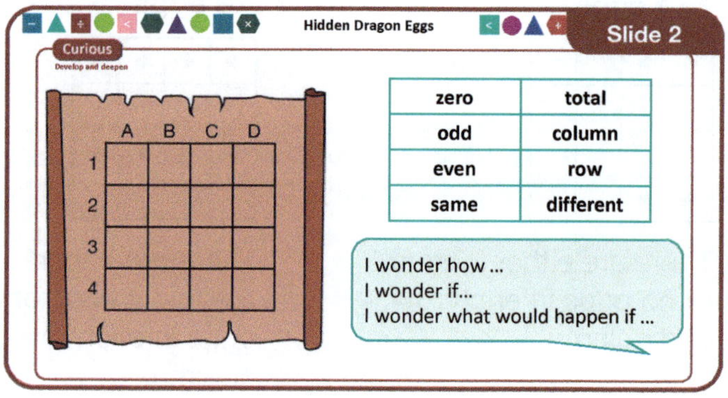

zero	**total**
odd	**column**
even	**row**
same	**different**

Possible lines of enquiry:

- I wonder if it is possible to make each column and row the same number.

Possible lines of enquiry linked to the word bank:

- I wonder if it will work with a total of 4 / 8 /12 eggs.
- I wonder how many eggs the dragon could be hiding.
- I wonder how many possibilities there are.
- I wonder whether there can be zero eggs in a column or row.

Possible lines of enquiry that children may raise now, but should be held back until the **Create** task:

- I wonder if it is possible to do the same but with an odd number of eggs in each row or column.
- I wonder how the rules could be changed to create a new challenge.
- I wonder what would happen if the size and shape of the grid was changed.

Consider whether / how to collate and display all the suggested questions, and whether / how to steer children towards selecting an appropriately challenging question.

Pupil Resources

Distribute the **Pupil Resources** for **Hidden Dragon Eggs** to the children to help support them in their exploration of the different tasks.

To provide additional flexibility in the classroom, each of the **Pupil Resources** is also provided in the **Teacher's Presentation** PowerPoint file (see slides 3–6), so you can display these at any time if you wish.

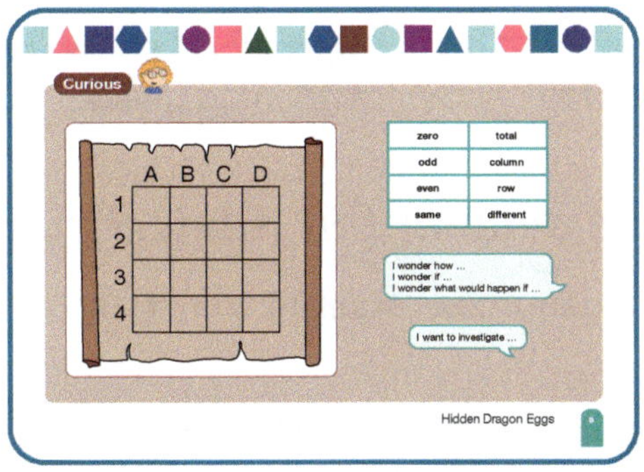

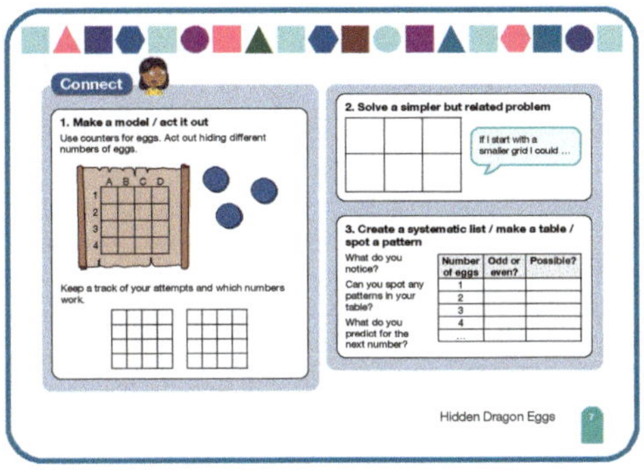

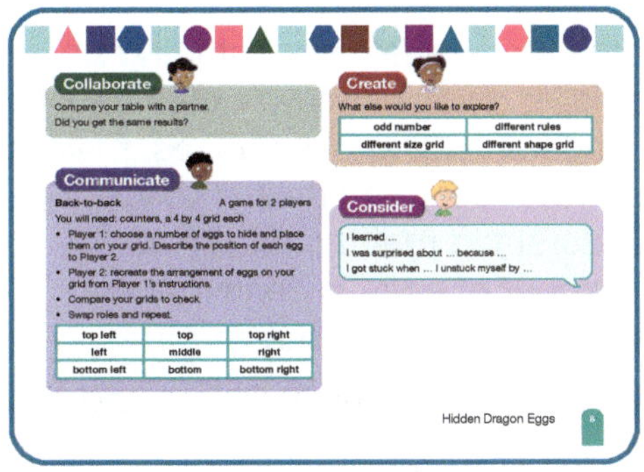

Three tasks involving different strategies are suggested for children to make connections.

1. Make a model / act it out

Additional resources required: counters

Before using any kind of systematic approach, children would benefit from acting out the problem and getting a feel for the parameters of the task.

Slides 7 and 8

Print slide 7 and distribute to children. Display slide 8.

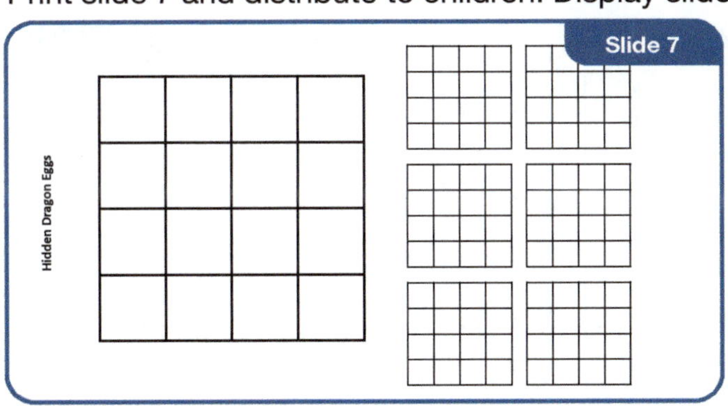

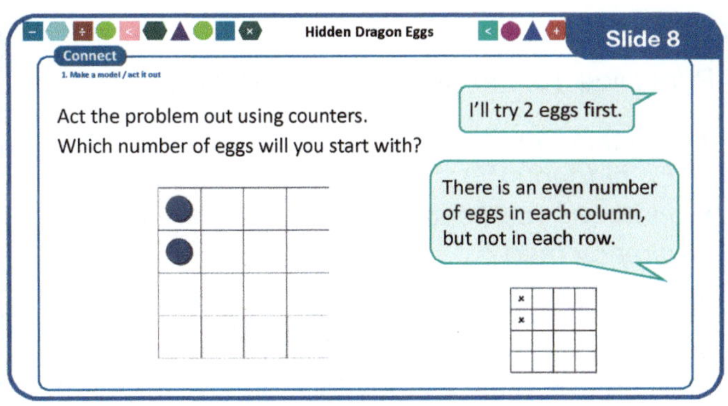

Ask children to use concrete resources, such as counters, to arrange different numbers of eggs on the grid.

As there are up to 16 totals to investigate, some children may need support to focus on the task. Support them by:

- Starting small. *Is it possible with 1, 2 or 3 eggs?*
- Building on success. *Great! You have found that it is possible to position ___ eggs. Is that the only possible arrangement?*
- Providing certainty. *I have found it is possible to find a way of arranging ____ eggs. Can you find a way?*

The smaller 4 by 4 grids are included on the printable slide so that children can track attempts and solutions and to encourage them to check that they aren't replicating a solution.

2. Solve a simpler but related problem

Additional resources required: counters

Slide 9

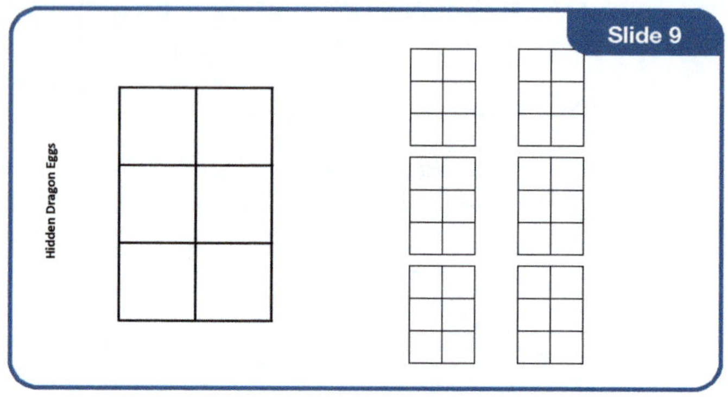

Print slide 9 and distribute to children. This provides a smaller 2 by 3 grid.

Some children may benefit from starting by exploring smaller grids with fewer counters, to gain experience of finding possibilities, making observations and recording their findings. Make sure they understand that the rule stays the same: the number of eggs in each row and column must be even (including zero).

Children could explore using 1, 2, 3, 4 and 5 counters.

- 1 counter is not possible as one row or column will have only one in.
- 2 or 4 counters work in lots of ways. You have to make sure each row or column has either 0 or 2.
- 3 and 5 are not possible as at least one row or column would have one counter in.

Children may make links between the smaller frame and the larger frame from the main task.

- *What did you notice so far? I wonder if it will be the same on the larger frame.*
- *Which arrangements worked on the smaller frame? Could you try this on the larger frame?*
- *Which arrangements did not work on the smaller frame? Will that be the same for the larger frame?*

Once children are more familiar with exploring the smaller grid, they can move on to the 4 by 4 grid.

3. Create a systematic list / make a table / spot a pattern

Slide 10

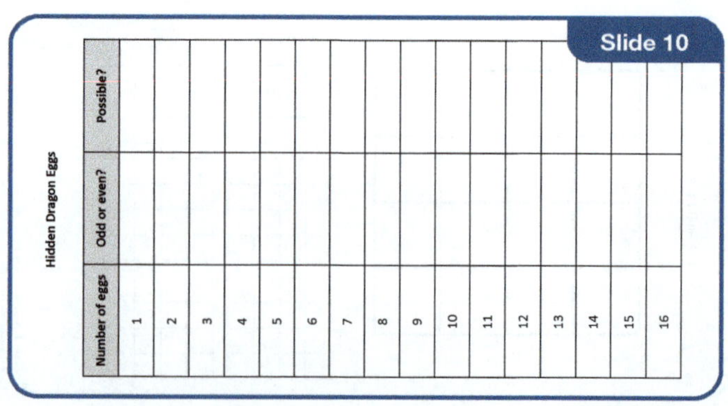

Print slide 10 and distribute to children.

- *What is the largest number of eggs that can be hidden? What about the smallest?*
- *How will you keep track of what you have already tried?*

To make sure children have explored every total, a systematic approach is needed. Ask children to fill out the table as they try different numbers.

- *Can you predict whether the next number will be possible or not?*

Slide 11

Display slide 11 which shows the completed table. Ask children to compare their lists. Did they get the same results?

- *Can you see any patterns?*
- *Is it possible to hide every even number of eggs between 1 and 16?*
- *Is it possible for any odd numbers?*
- *Do any of these results surprise you?*

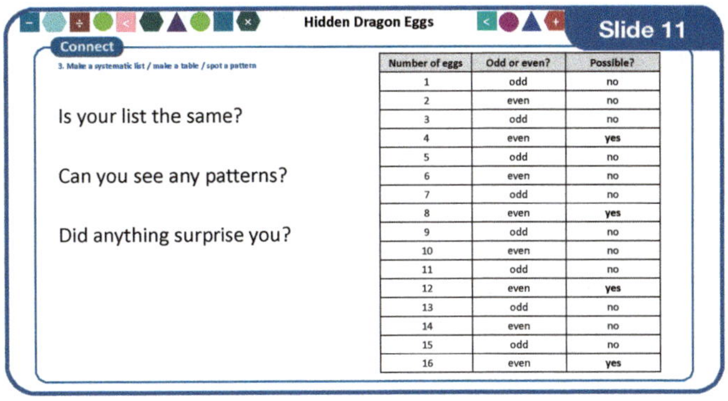

Number of eggs	Odd or even?	Possible?
1	odd	no
2	even	no
3	odd	no
4	even	**yes**
5	odd	no
6	even	no
7	odd	no
8	even	**yes**
9	odd	no
10	even	no
11	odd	no
12	even	**yes**
13	odd	no
14	even	no
15	odd	no
16	even	**yes**

Refer to the **Collaborate** task of the **Pupil Resource**.

This task provides an opportunity for children to work independently or in pairs to follow their chosen line(s) of enquiry. Each child or pair can build their models and then compare with others how they have arranged a certain number of eggs, (for example, 11) and discuss whether it is possible to arrange the eggs in even rows and columns or not.

- *Did you arrange your eggs (counters) in the same way?*
- *Do you agree on which numbers of eggs are possible and which aren't? Does your recording table look the same?*

Children will play the game in the **Communicate** task together, which provides them with the opportunity to rehearse their mathematical vocabulary and requires careful and precise speaking and listening skills.

Refer to the **Communicate** task of the **Pupil Resource**.

Slide 12

Display slide 12.

This task provides an opportunity for the children to rehearse positional language by writing or verbally giving instructions of how to use the map to find the eggs.

How to play:

Each player has a map with the 4 by 4 grid and some counters. Ask pairs to sit back-to-back.

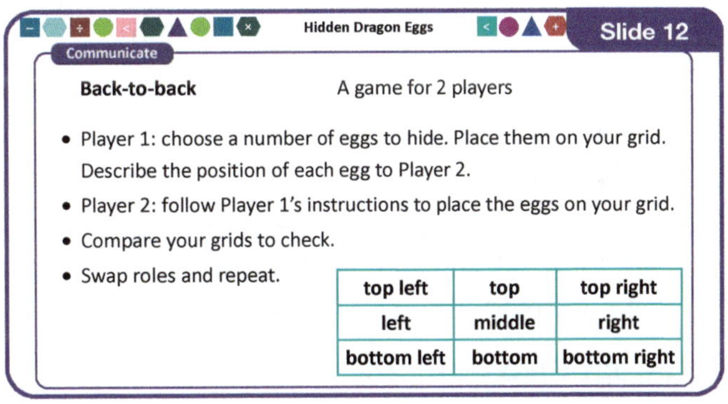

- Player 1 chooses a number of eggs to hide and places them on their grid.
- They describe the position of each egg to Player 2.
- Player 2 recreates the arrangement of eggs on their grid from Player 1's instructions.
- Players compare their grids to check.
- They swap roles and repeat.

Children may use the vocabulary bank provided to scaffold their descriptions.

top left	**top**	**top right**
left	**middle**	**right**
bottom left	**bottom**	**bottom right**

⚠ Children may need support with identifying the 'top left' or 'bottom right' dot using a combination of positional language.

Create

Refer to the **Create** task of the **Pupil Resource**.

Slide 13

Display slide 13.

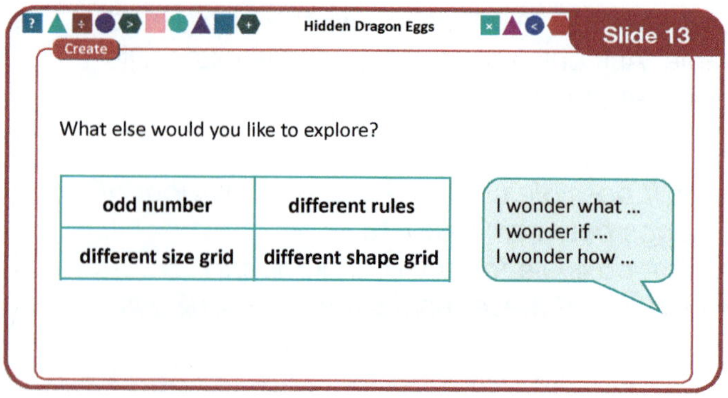

- *Were there any questions asked at the start of the task that you now want to think about?*
- *Is there anything new that you are wondering now?*

Children can use the word bank to prompt further questions of their own.

odd number	**different rules**
different size grid	**different shape grid**

- I wonder if it is possible to do the same but with an odd number of eggs in each row or column.
- I wonder how the rules could be changed to create a new challenge.
- I wonder what would happen if the size and shape of the grid was changed.

Consider

Refer to the **Consider** task of the **Pupil Resource**.

Slide 14

Display slide 14.

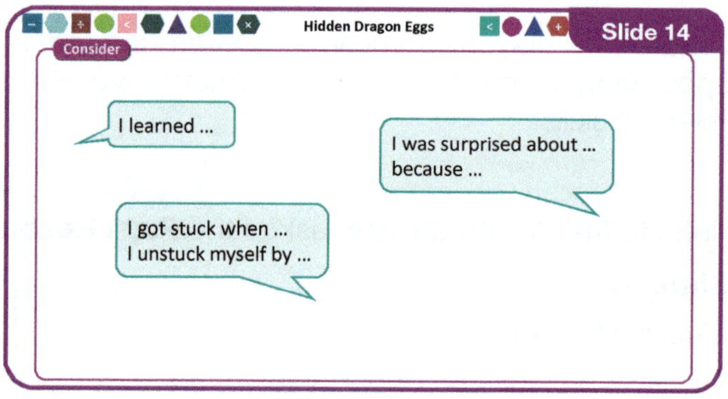

Encourage children to think about the strategies that they have used. The prompts below may help get children started.

Cognitive learning domain

- The trickiest part of solving my question was …
- I learned …

Affective domain

- The fun bit of my learning was … It was fun because …
- I said 'aha' when … I said it because …
- I was surprised about … because …

Metacognitive domain

- I got stuck when … I unstuck myself by …
- I used … to help me. It was helpful because …
- Next time, I will …

Answers

Numbers	Comment	Diagram
1, 3, 5, 7, 9, 11, 13, 15	No odd numbers are possible, as at least one column and / or row would have an odd number.	
2	Not possible as at least two rows or columns would have one counter.	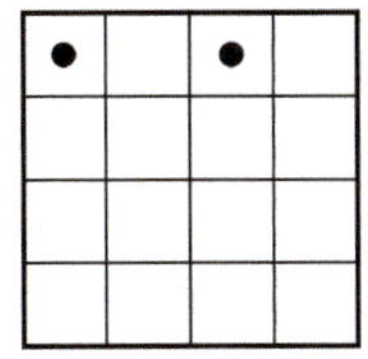
4	Numerous different combinations possible including:	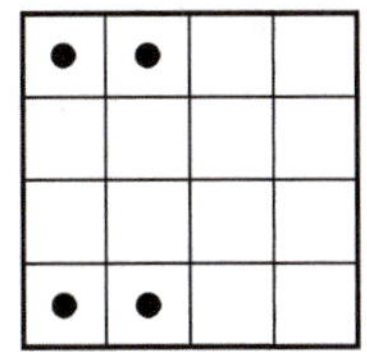
6	Not possible – there is always at least one row or column with an odd number of counters.	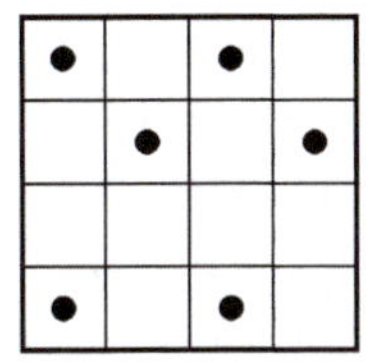
8	Numerous different combinations including:	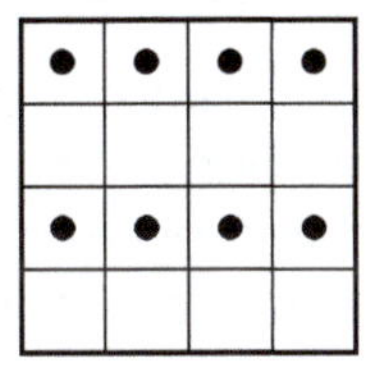
10	Not possible – there are always at least two rows or columns with an odd number of counters.	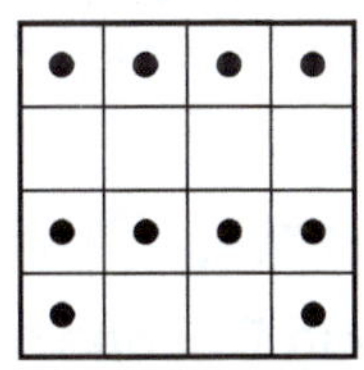
12	Numerous different combinations including:	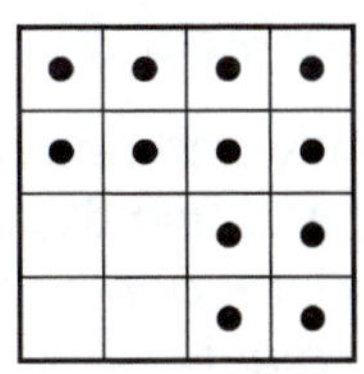
14	Not possible as the two empty spaces always mean some of the rows or columns would have three counters.	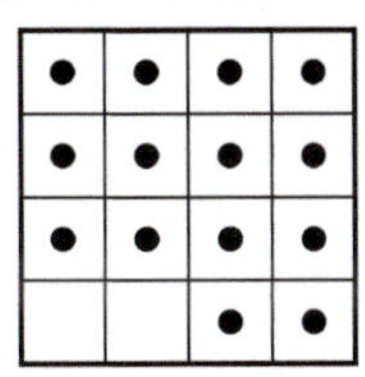
16	Possible.	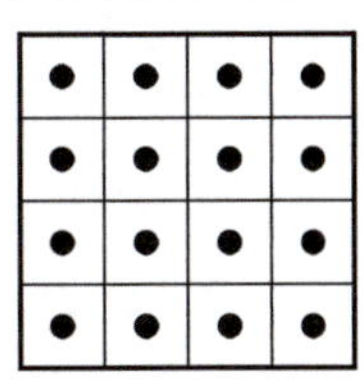

Dots and Dashes

Year 1

Introduction

Dots and Dashes provides children with an opportunity to explore a different representation of the number system. They identify, and then use, patterns together with their counting skills to work out what each diagram stands for and explain their reasoning.

Once children have explored the diagrams for the numbers 4 and 8, they explore the connections between the dots and dashes and use what they have noticed to represent other numbers.

Throughout the task, children are supported to predict, and justify their predictions, through the use of speaking frames.

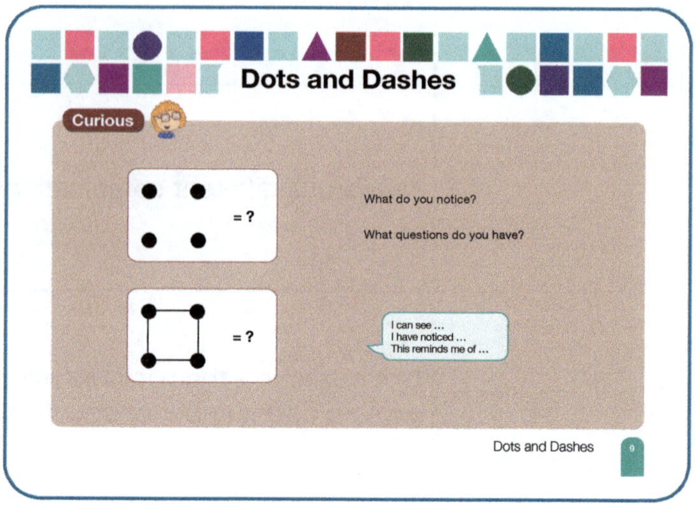

This task promotes greater-depth learning in a range of ways, supporting children to identify and articulate underlying mathematical structures. For example, in the **Collaborate** stage, greater depth is promoted by supporting children to be polite and encouraging to group members, and to offer their own ideas and accept it if these are rejected by the group.

This **Teacher Guide** is supported by photocopiable **Pupil Resources** and **Teacher's Presentation** PowerPoint slides for whole-class or group display.

National Curriculum

Dots and Dashes covers these Year 1 National Curriculum objectives.

Number and place value

- Identify and represent numbers using objects and pictorial representations including the number line, and use the language of: equal to, more than, less than (fewer), most, least.
- *Pupils recognise and create repeating patterns with objects and with shapes.* [non-statutory guidance]

The 6C cycle

The grid on the right shows the relative weighting given to each of the 6 Cs in **Dots and Dashes.**

Curious	★★★
Connect	★★
Collaborate	★★★
Communicate	★★★
Create	★
Consider	★

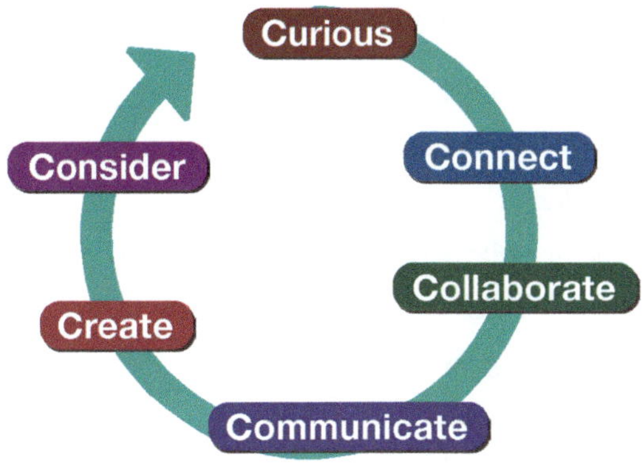

Problem-solving heuristics

The grid shows which problem-solving strategies children are likely to use when investigating the Connect tasks in **Dots and Dashes.** These strategies are further exemplified in the Connect stage below (pages 45–48).

Draw a diagram	✓
Make a table / spot a pattern	✓
Make a model / act it out	
Try then improve	
Create a systematic list	
Solve a simpler but related problem	✓
Work backwards	✓

Key mathematical ideas

The counting system used in **Dots and Dashes** is similar to tally marks where things are counted and then represented in bundles of 5. The system used in this task, Tukey tallying, is named after John Tukey. Numbers from 1 to 4 are represented by dots in the arrangement of a square. The numbers 5 to 10 are then represented by lines that join the dots, firstly by working around the outside of the square and then joining diagonals.

1	2	3	4	5	6	7	8	9	10

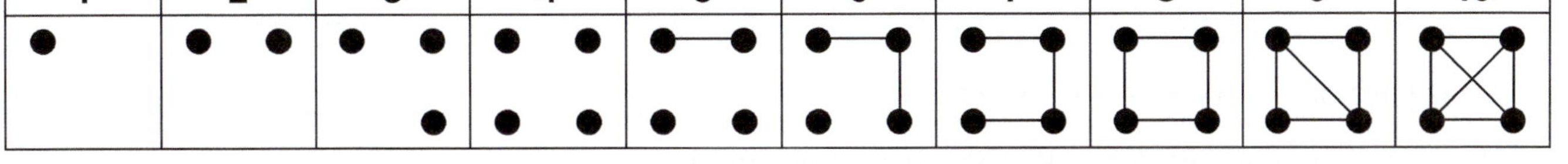

Children will benefit most from the task if they are confident with one-to-one correspondence, and can accurately use language such as more than, less than, fewer than and difference.

Answers to the tasks (the diagrams for numbers 1–20) are given on page 50.

Teaching guidance

Teaching guidance is given below.

The **Teacher's Presentation** PowerPoint file has slides specifically mentioned in this teaching guidance. Some of the slides include notes (within the PowerPoint file) providing 'stumble support' suggestions, that may help you to support children in their problem-solving thinking and enable access for children who are less confident.

Awe and wonder

Reveal slides 1–2 one at a time, allowing time for discussion and observations.

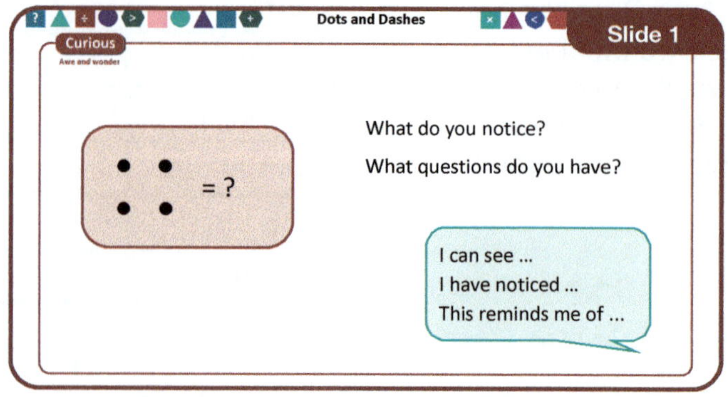

Slide 1

Allow children to share what they are noticing or wondering.

- I can see four dots.
- It reminds me of a dice.
- I have noticed that there is an equals sign and a question mark.

Slide 2

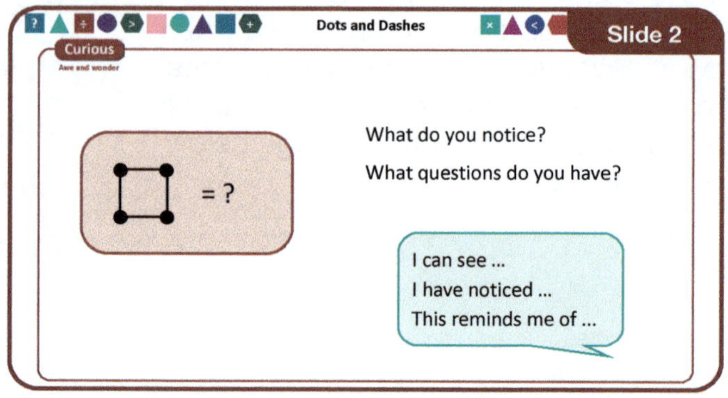

- I can see 4 dots and 4 lines.
- I have noticed that it is similar to the 4 dots. But this one has 4 lines as well.

Encourage children to compare the two diagrams.

- I have noticed that the other one has fewer parts because the dots aren't joined with lines.

Develop and deepen

Slide 3

Display slide 3.

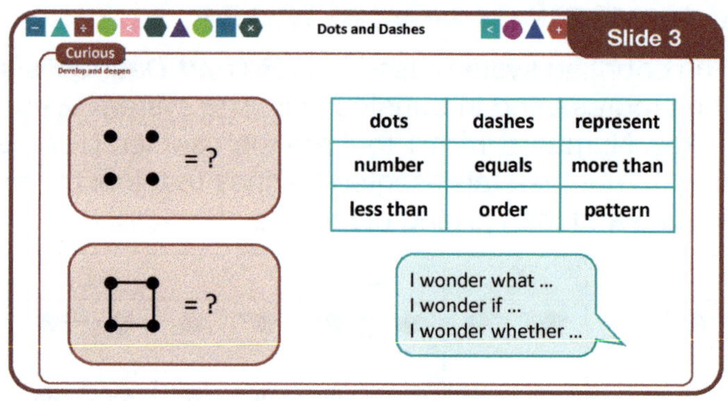

Encourage children to use the words in the word bank to develop and deepen their thoughts and questions.

dots	**dashes**	**represent**
number	**equals**	**more than**
less than	**order**	**pattern**

Possible lines of enquiry linked to the word bank:

- I wonder what number each question mark represents.
- I wonder if I can find any patterns.
- I wonder which diagram is worth more than the other.
- I wonder if the more dots and dashes, the higher the number.
- I wonder if I could represent other numbers and put them in order.

Possible lines of enquiry that children may raise now, but should be held back until the **Create** task:

- I wonder if I could draw two-digit numbers.
- I wonder what the number 100 would look like.

Consider whether / how to collate and display all the suggested questions, and whether / how to steer children towards selecting an appropriately challenging question.

At this point, tell children what number each diagram represents.

The first diagram represents a 4, the second an 8.

- *Can you see how each number is represented?*
- *How do the dots and dashes link to 4 and 8?*

Pupil Resources

Distribute the **Pupil Resources** for **Dots and Dashes** to the children to help support them in their exploration of the different tasks.

To provide additional flexibility in the classroom, each of the **Pupil Resources** is also provided in the **Teacher's Presentation** PowerPoint file (see slides 4–7), so you can display these at any time if you wish.

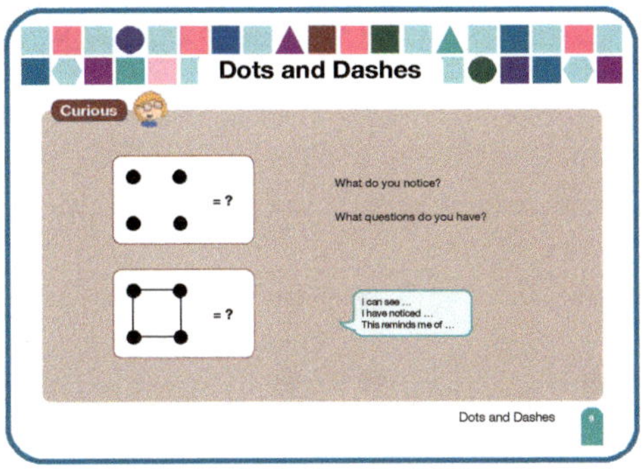

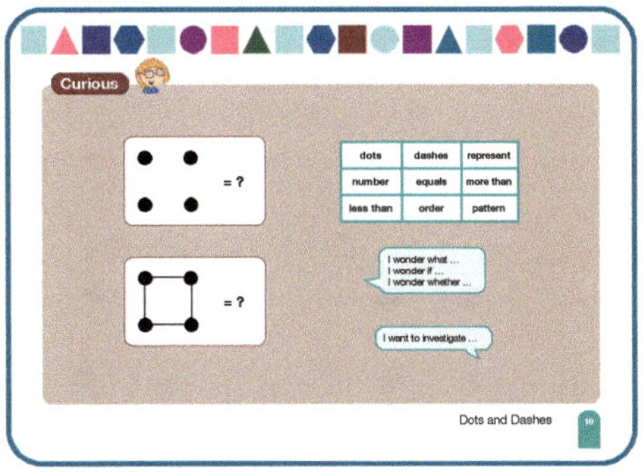

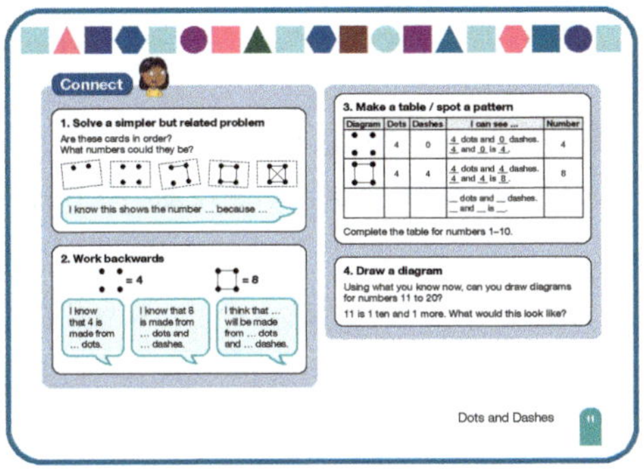

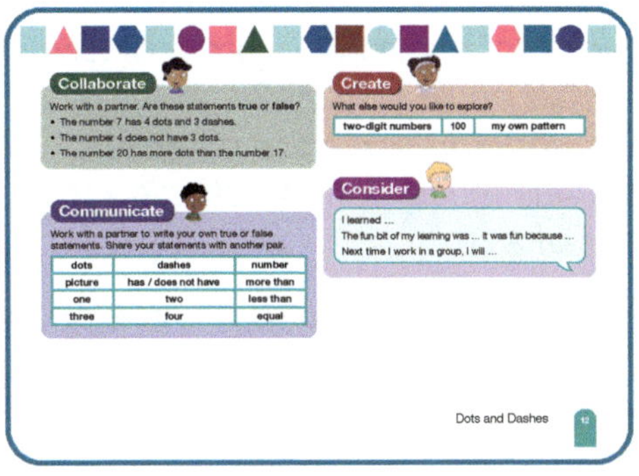

Four tasks involving different strategies are suggested for children to make connections.

Slide 8

Print slide 8 and cut out the cards. Distribute these cards to the children as described in the **Connect** tasks. The blank cards can be used by children to draw diagrams for numbers 11–20, or other numbers that they wish to explore in the **Create** stage.

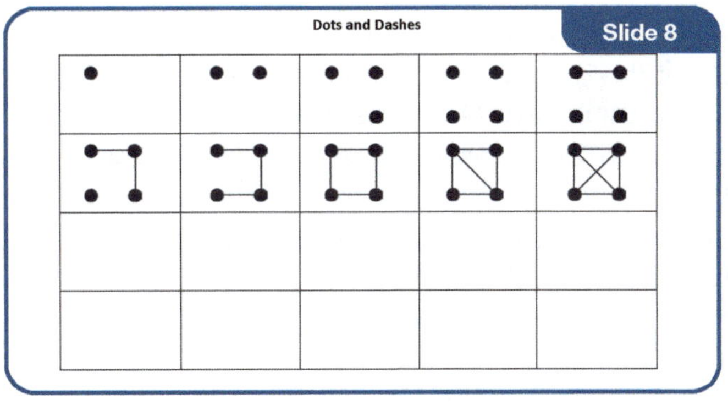

1. *Solve a simpler but related problem*

Slide 9

Display slide 9.

Give children the cut-out cards from slide 8 showing the even numbers only.

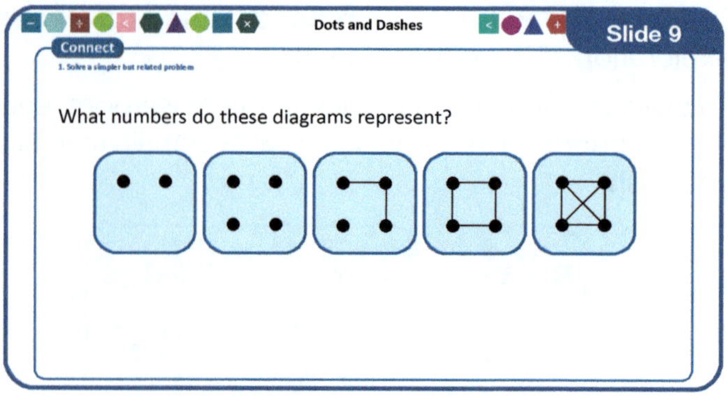

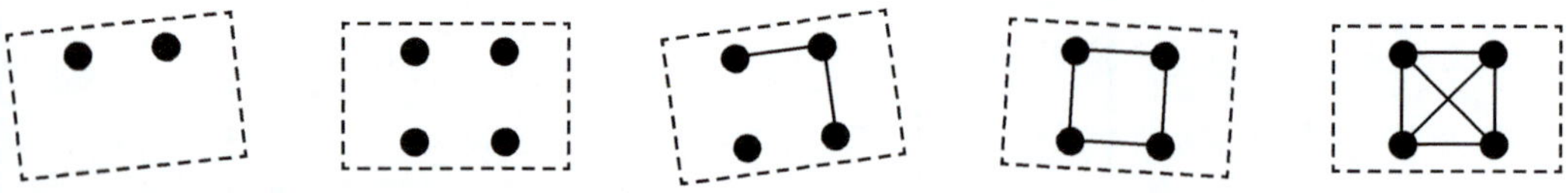

Ask them to arrange their cards in the same order as shown on the slide.

- *Do you think that these are in order of size?*
- *What numbers could they be?*
- *Which diagram represents the number 6? Number 2? How do you know?*
- I know that 4 is made from 4 dots, so I think that the 2 dots represent the number 2. I think these are in order of size.

Give children the remaining cut-out cards and ask them to order the diagrams from smallest to largest value.

2. *Work backwards*

Slide 10

Encourage children to consider what they know about the diagrams for the numbers 4 and 8 and to think about how this knowledge can help them to work out what other numbers would look like.

Give children the cut-out cards for numbers 1, 2, 4, 5, 6, 8, 9 and 10, and two blank cards. Ask them to work out which digit each card represents, and then draw the diagrams for 3 and 7 on the blank cards. For those who require a little more support, give them the cut-out cards for all numbers 1–10.

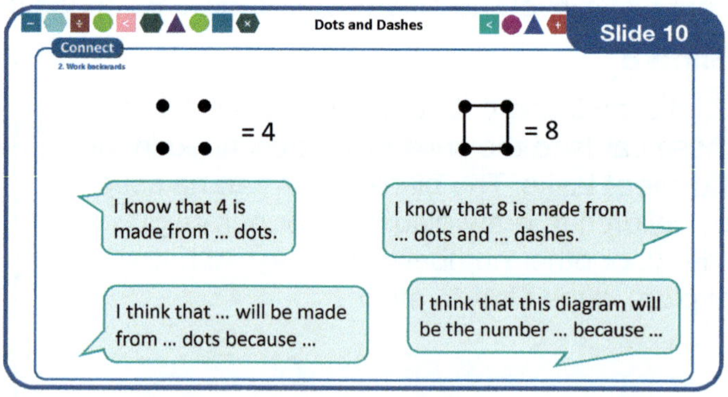

Display slide 10 which provides useful sentence stems.

3. Make a table / spot a pattern

Slide 11

Print slide 11 and distribute to children. Ask them to use what they notice about the diagrams for 4 and 8 to fill in the table. They should draw the diagram in the first column, and then fill in the detail for the other columns.

Model filling in the table for the number 5.

- *How can we work out what 5 would look like? What do we already know?*

Once children have filled in some of the rows in the table ask:

- *Have you noticed any patterns from your table or your drawings?*
- It goes 1 dot, 2 dots, 3 dots, 4 dots and then 4 dots and 1 dash.
- When you get 4 dots, you then go up in dashes. 4 dots and 1 dash, then 4 dots and 2 dashes.
- If you add the number of dots and then number of dashes, this gives you the number. So 4 dots and 1 dash is 5.

Diagram	Dots	Dashes	I can see ...	Number
	1	0	1 dot and 0 dashes. 1 and 0 is 1.	1
	2	0	2 dots and 0 dashes. 2 and 0 is 2.	2
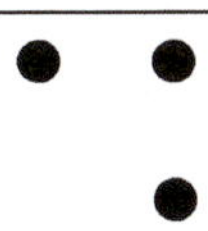	3	0	3 dots and 0 dashes. 3 and 0 is 3.	3
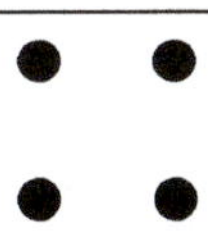	4	0	4 dots and 0 dashes. 4 and 0 is 4.	4
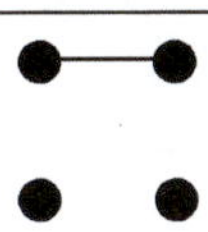	4	1	4 dots and 1 dash. 4 and 1 is 5.	5
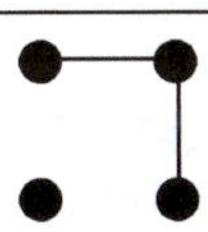	4	2	4 dots and 2 dashes. 4 and 2 is 6.	6
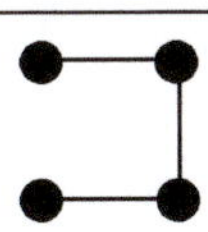	4	3	4 dots and 3 dashes. 4 and 3 is 7.	7
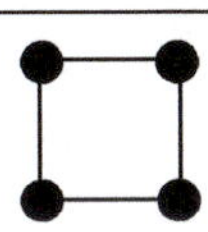	4	4	4 dots and 4 dashes. 4 and 4 is 8.	8
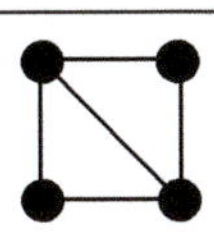	4	5	4 dots and 5 dashes. 4 and 5 is 9.	9
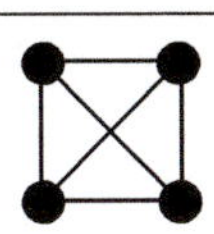	4	6	4 dots and 6 dashes. 4 and 6 is 10.	10

Completed table from slide 11

Connect

4. Draw a diagram

Give children some of the blank cut-out cards from slide 8.

This task could also be used together with the **Create** task.

- *Using what you know now, can you draw diagrams for numbers 11 to 20?*
- *What do you know about the number 11?*
- 11 is 1 ten and 1 more.
- *What would this look like?*
- I think it will be like this: the diagram for 10, with an extra dot.

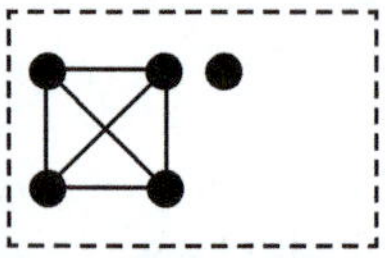

- *What do you think 20 would look like?*
- 20 is 10 and another 10. I think it will be made with 2 of the 10 pictures.

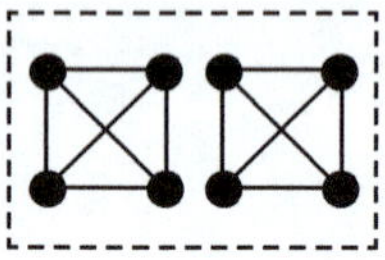

Collaborate

Refer to the **Collaborate** task of the **Pupil Resource**.

Slide 12

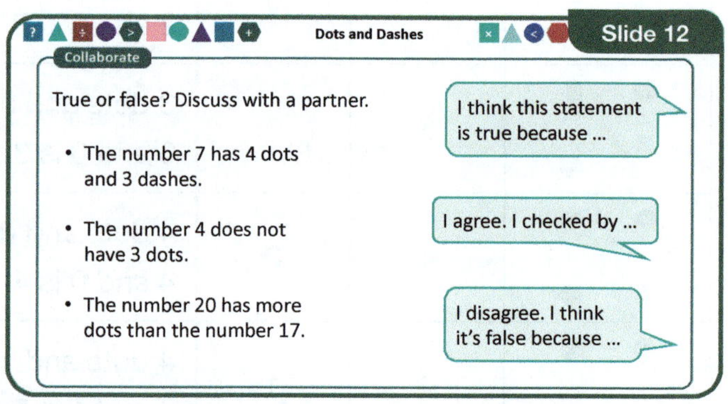

Display slide 12.

This task provides an opportunity for children to decide whether statements about the diagrams for the numbers that they have explored are true or false.

- The number 7 has 4 dots and 3 dashes. [True]
- The number 4 does not have 3 dots. [True]
- The number 20 has more dots than the number 17. [False]

Ask children to take it in turns to read out the statements and then discuss whether they agree or disagree. Encourage them to give reasons for their decisions using the sentence stems on the slide.

Refer to the Communicate task of the **Pupil Resource**.

Slide 13

Display slide 13.

This task provides an opportunity for the children to write their own true or false statements about the diagrams for the numbers that they have explored. Each statement could be written on a slip of paper and then swapped with other pairs.

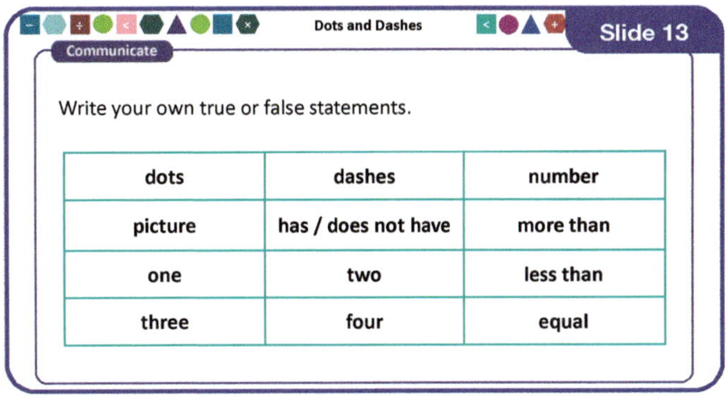

Children can use the word bank to help them.

dots	**dashes**	**number**
picture	**one**	**two**
three	**four**	**has / does not have**
more than	**less than**	**equal**

Refer to the Create task of the **Pupil Resource**.

Slide 14

Display slide 14.

- *Were there any questions asked at the start of the task that you now want to think about?*
- *Is there anything new that you are wondering now?*

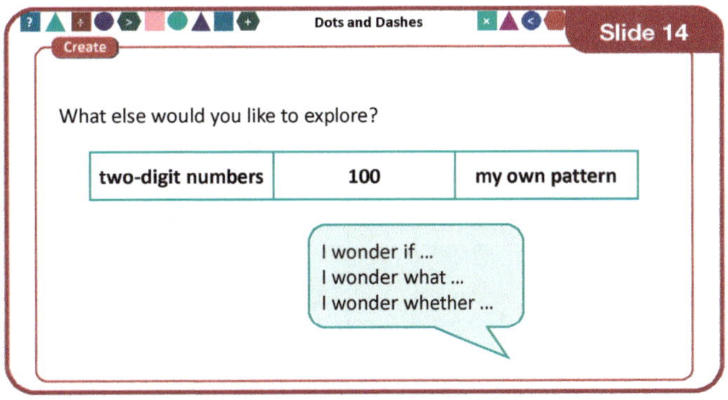

Children can use the word bank to prompt further questions of their own.

two-digit numbers	**100**	**my own pattern**

- I wonder if I could draw two-digit numbers.
- I wonder what the number 100 would look like.
- I wonder whether I could create my own pattern for counting.

Consider

Refer to the **Consider** task of the **Pupil Resource**.

Slide 15

Display slide 15.

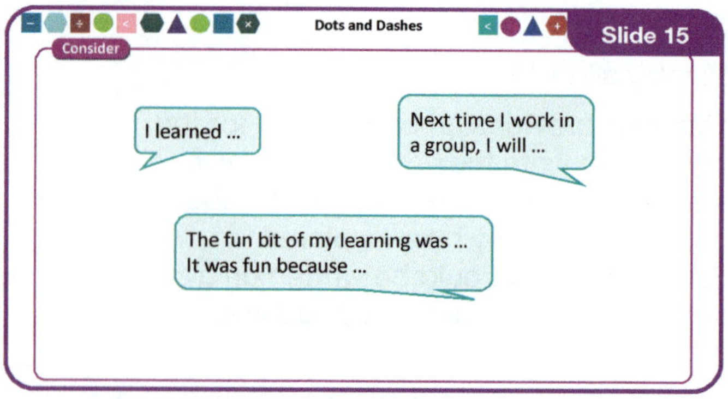

Encourage children to think about the strategies that they have used. The prompts below may help get children started.

Cognitive learning domain

- I learned …
- A question that I can answer now that I have completed this task is …

Affective domain

- The fun bit of my learning was … It was fun because …
- I said 'aha' when … I said it because …

Metacognitive domain

- I think that I worked … with other people. I think this because …
- Next time I work in a group, I will …

Answers

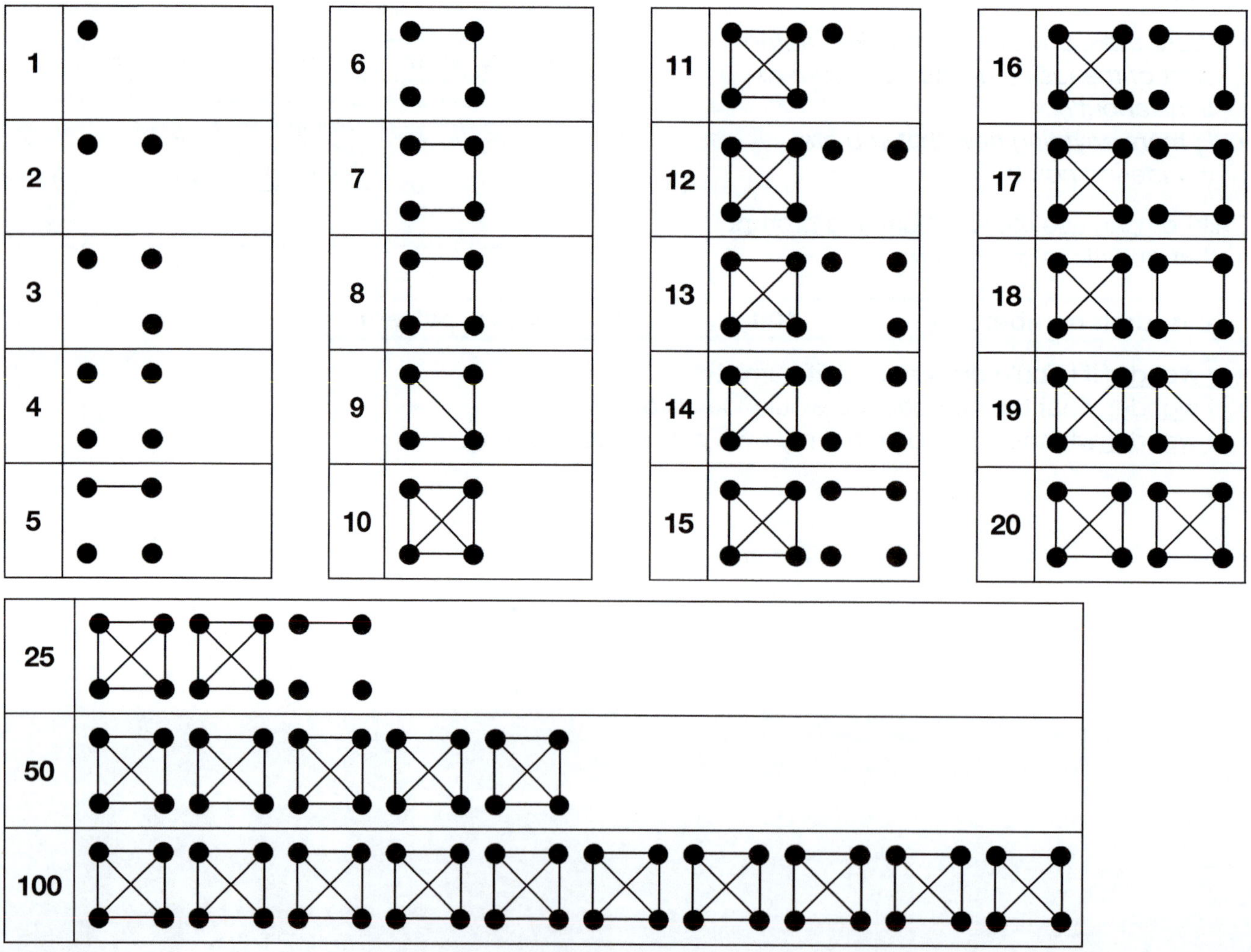

Money Maze

Year 1 Year 2

Introduction

Money Maze provides children with the opportunity to rehearse and apply adding one-digit numbers within the engaging context of using money. Additional complexity is added through the use of multiple coins in each grid square and the possibility of a number of different lines of enquiry. Children may choose to calculate the value of each square before they start and assess whether this is an efficient way to work rather than adding all of the coins each time.

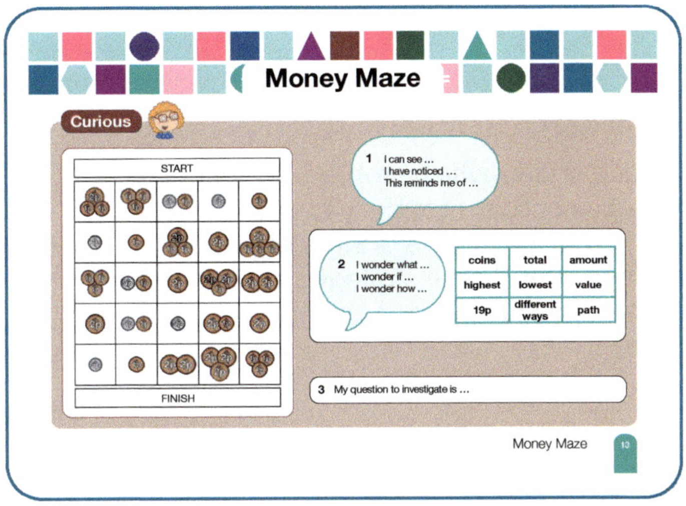

The task allows children to consider different resources to help them with their calculations, to keep track of their totals and to also provide proof. They use mental methods, jottings and concrete resources and consider in depth the suitability and efficiency of the strategies that they choose.

This task promotes greater-depth learning in a range of ways, supporting children to identify and articulate underlying mathematical structures. For example, in the **Connect** stage, greater depth is promoted by supporting children to choose an appropriate strategy and explain their choice, and understand that sometimes more than one answer is possible.

This **Teacher Guide** is supported by photocopiable **Pupil Resources** and **Teacher's Presentation** PowerPoint slides for whole-class or group display.

National Curriculum

Money Maze covers these Year 1 and Year 2 National Curriculum objectives.

Year 1 Addition and subtraction

- Add and subtract one-digit and two-digit numbers to 20, including zero.

Year 1 Measurement

- Recognise and know the value of different denominations of coins and notes.

Year 2 Addition and subtraction

- Add and subtract numbers using concrete objects, pictorial representations, and mentally, including: a two-digit number and ones; a two-digit number and tens; two two-digit numbers; adding three one-digit numbers.

Year 2 Measurement

- Find different combinations of coins that equal the same amounts of money.
- Solve simple problems in a practical context involving addition and subtraction of money of the same unit, including giving change.

The 6C cycle

The grid on the right shows the relative weighting given to each of the 6 Cs in **Money Maze.**

Curious	★★
Connect	★★★
Collaborate	★★
Communicate	★
Create	★
Consider	★★

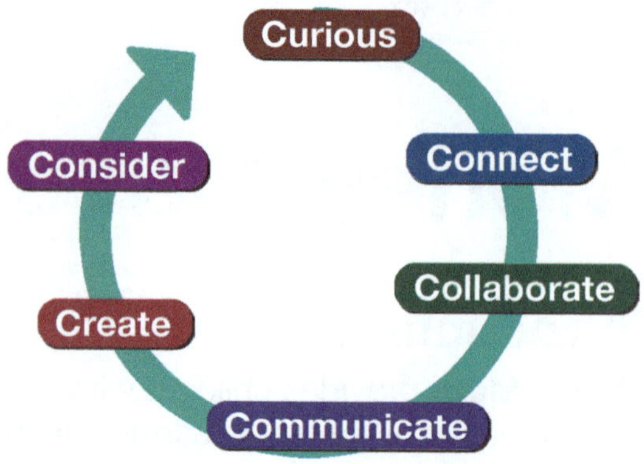

Problem-solving heuristics

The grid shows which problem-solving strategies children are likely to use when investigating the **Connect** tasks in **Money Maze**. These strategies are further exemplified in the **Connect** stage below (pages 54–57).

Draw a diagram	
Make a table / spot a pattern	
Make a model / act it out	✓
Try then improve	✓
Create a systematic list	✓
Solve a simpler but related problem	✓
Work backwards	✓

Key mathematical ideas

The aim of this task is to provide children with opportunities to rehearse and apply strategies for adding one-digit numbers. Children will get the most out of this task if they are secure in different methods for addition and subtraction.

Useful strategies could include:

- using near doubles: 5p + 6p = double 5p + 1p or double 6p – 1p
- looking for complements to 5: 3p + 4p + 2p could be rearranged to 3p + 2p + 4p
- looking for complements to 10: 3p + 5p + 7p could be rearranged to 3p + 7p + 5p
- regrouping to count on: 6p + 8p, regroup 6p into 2p and 4p, 8p + 2p + 4p
- regrouping to subtract: 15p – 8p, regroup 8p into 5p and 3p, 15p – 5p – 3p
- regrouping to subtract: 15p – 8p, regroup 15p into 10p and 5p, 10p – 8p + 5p
- counting on to subtract: 15p – 8p, counting on 2p from 8p to 10p and then counting on another 5p.

This task also gives children the opportunity to rehearse their knowledge of coins and how totals can be made with different combinations of coins.

Answers to some of the tasks are given on page 59, but other answers will vary depending on the lines of enquiry chosen.

Teaching guidance

If you are using the **Express Pathway**, please turn to pages 54–59.

Teaching guidance for the **Full 6C Pathway** is given below.

The **Teacher's Presentation** PowerPoint file has slides specifically mentioned in this teaching guidance. Some of the slides include notes (within the PowerPoint file) providing 'stumble support' suggestions, that may help you to support children in their problem-solving thinking and enable access for children who are less confident.

Awe and wonder

Slide 1

Display slide 1.

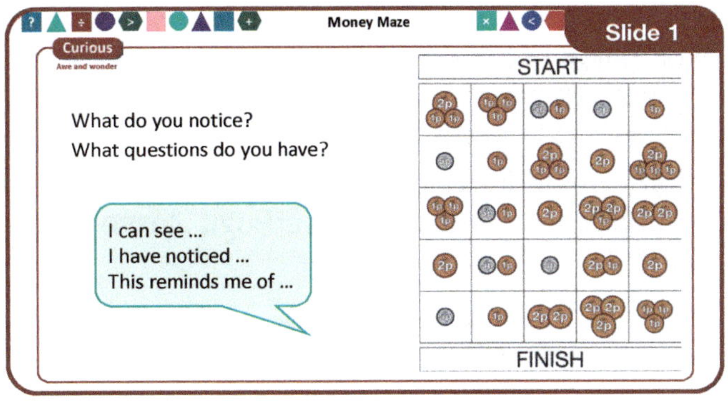

- I can see lots of coins.
- I can see a grid with 25 boxes.
- I have noticed that lots of the boxes have more than 1 coin in them.
- I have noticed that there are 1p, 2p and 5p coins.
- I have noticed that there is a start and finish.
- It reminds me of a treasure map a bit with the grid.

At this point, explain that you start from the top and choose a path to the finish line. Tell children that they can start on any square in the top row and finish on any square in the bottom row. They can move down, left, right and diagonally.

Develop and deepen

Encourage children to use the words in the word bank to develop and deepen their thoughts and questions.

coins	**total**	**amount**
highest	**lowest**	**value**
19p	**different ways**	**path**

Possible lines of enquiry linked to the word bank:

- I wonder what the total of each box is.
- I wonder what the highest / lowest value path to the finish line is.
- I wonder what the highest total I can make with 8 moves is.
- I wonder if two paths will have the same total.
- I wonder how many different ways there are to make a total of 19p.
- I wonder how many ways there are to make a different target.

Possible lines of enquiry that children may raise now, but should be held back until the **Create** task:

- I wonder what would happen if I changed the start and finish line and went across the board instead of down. What would the highest total be?

Consider whether / how to collate and display all the suggested questions, and whether / how to steer children towards selecting an appropriately challenging question.

At this point, confirm that you would like children to try to find paths that make a total of 19p.

Pupil Resources for Full and Express Pathways

Distribute the **Pupil Resources** for **Money Maze** to the children to help support them in their exploration of the different tasks. Use either of the **Curious Pupil Resources** depending on whether you are following the **Full Pathway** or the **Express Pathway**. The **Pupil Resources** for the remaining 5Cs are common to both pathways.

To provide additional flexibility in the classroom, each of the **Pupil Resources** is also provided in the **Teacher's Presentation** PowerPoint file (see slides 2–5), so you can display these at any time if you wish.

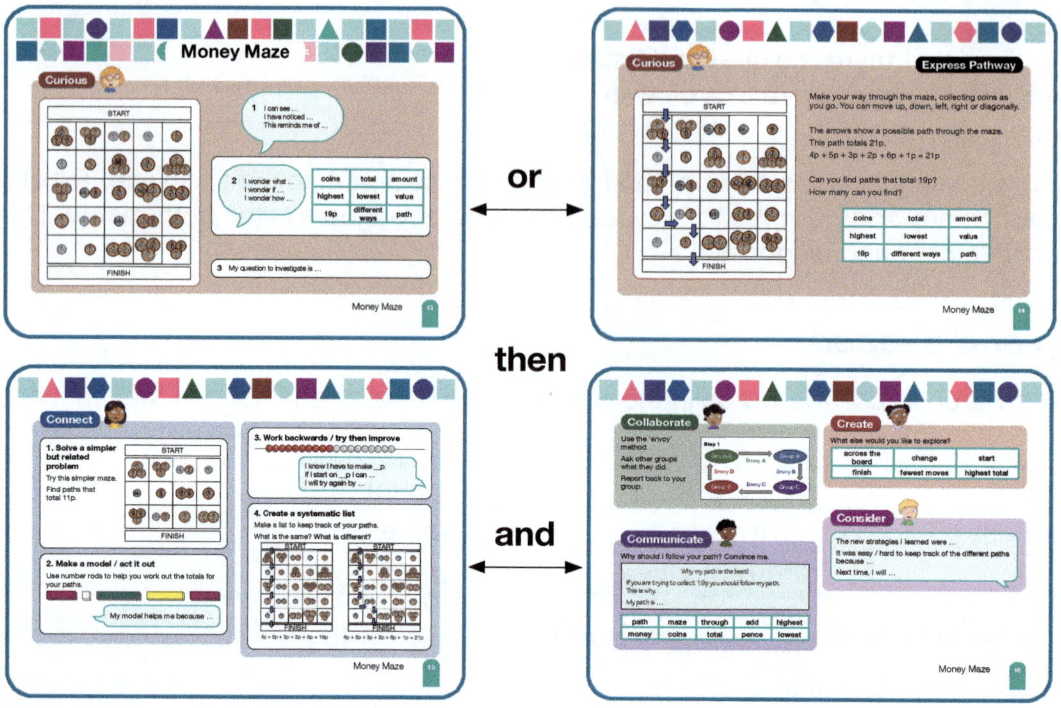

Express Pathway

If you are following the **Express Pathway**, spend a few minutes discussing the **Curious Pupil Resource** with the children. (This is slide 3, so you could display it if you wish.) Use some of the prompts on the previous page to help.

Four tasks involving different strategies are suggested for children to make connections.

Both the main grid and the simpler grid are provided as printable slides (slides 6 and 7).

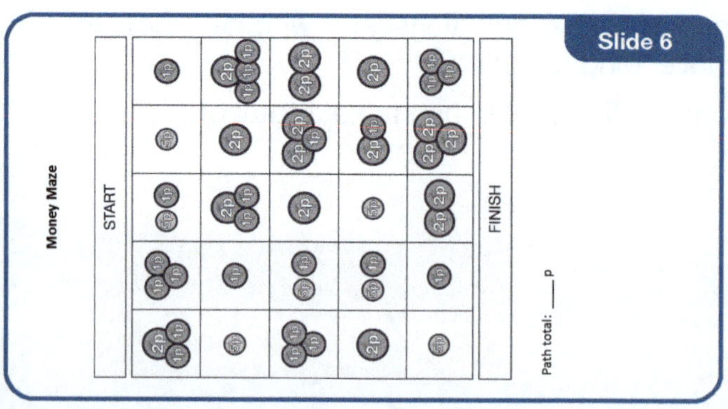

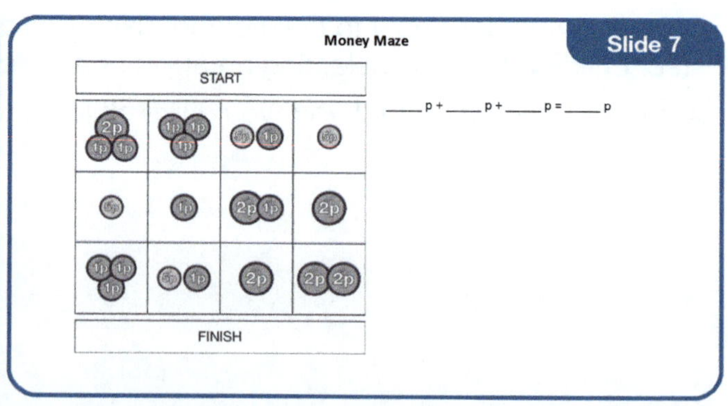

After printing, you may wish to laminate them for repeated use. Children can write the total in each box so that they do not have to work out the value each time they choose a different path.

1. Solve a simpler but related problem

Slide 7

Print slide 7 and distribute to children.

Before tackling the full problem, children may benefit from exploring a smaller maze with 12 squares instead of 25 and a smaller target number. Ask children to find a path that totals 11p.

Making a model, working backwards, try then improve and creating a systematic list, as detailed below, can be used as strategies to explore this simpler problem as well as for the full 5 by 5 grid.

2. Make a model / act it out

Additional resources required: number rods, number rod tracks, bead strings (optional)

Slide 8

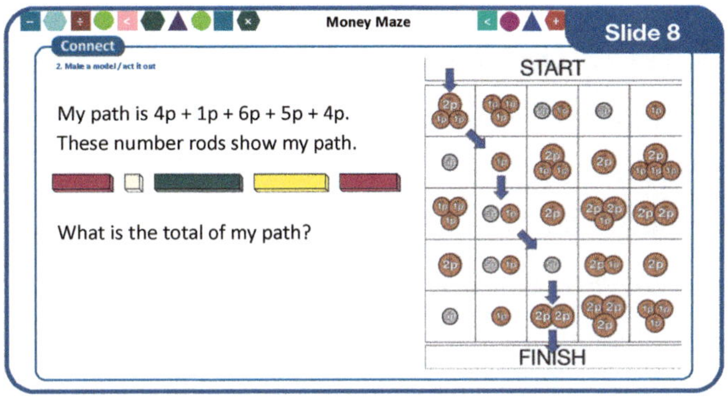

Display slide 8.

Children could make a model of the path they choose using concrete resources such as number rods. Encourage them to then use the resources to help them calculate.

Model this with slides 8–10. Explain to children that the arrows show one possible path.

- *My path is 4p then 1p then 6p then 5p then 4p. How do the number rods show this?*
- The 4-rod represents 4p so there are two 4-rods, a 1-rod, a 6-rod and a 5-rod.

Ask children to arrange number rods as shown on the slide.

- *What strategies could you use to add up these numbers? Can you rearrange the number rods to help you?*

Slide 9

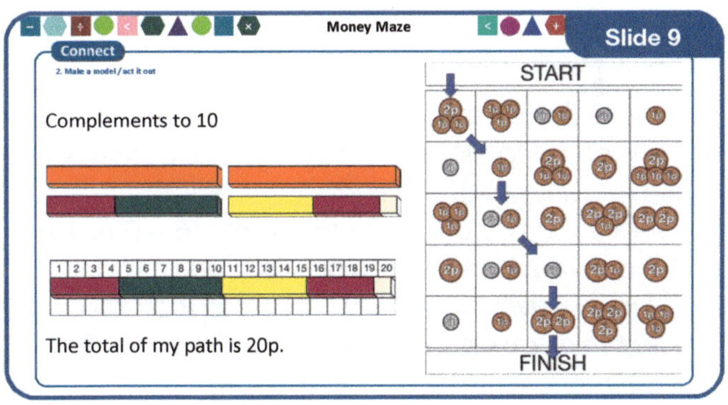

Display slide 9.

- *I have found ways of making 10. 4p and 6p is 10. 4p and 1p is 5p and another 5p makes 10p. I have 20p.*
- *I have rearranged the sum to 4p + 6p + 5p + 4p + 1p to make it easier.*

Slide 10

Display slide 10.

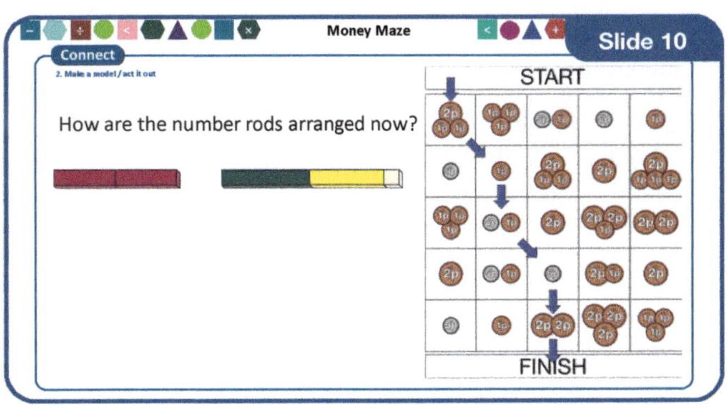

- *Can you see how I have rearranged the rods this time?*
- The two 4-rods are together.
- *Why do you think I put the 4-rods together?*

Explain that double 4 is a known fact and that it can be recalled quickly.

- *I have used doubles. Double 4 is 8. I can use the 6-rod, the 5-rod and 1-rod to make two 6s which is 12. 8p and 12p is 20p.*

Ask children which method they preferred.

Suggest that children make up the target number of 19p using number rods, and then compare this to a rod path that they are looking to calculate. This helps encourage estimating before they calculate. For example:

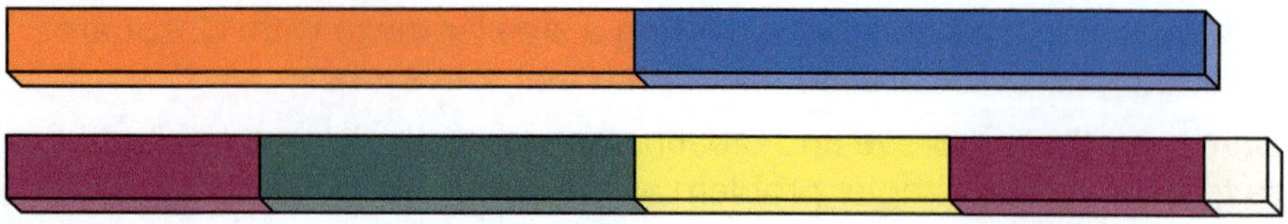

- *What does your model tell you before you do any adding?*
- I can see that my rods are longer than 19p. I can see that it is only a little bit longer.
- I know that my rods will equal 20p because I have 1 more than 19p. I will add them up to check.

3. Work backwards / try then improve

Additional resources required: bead strings

Slide 11

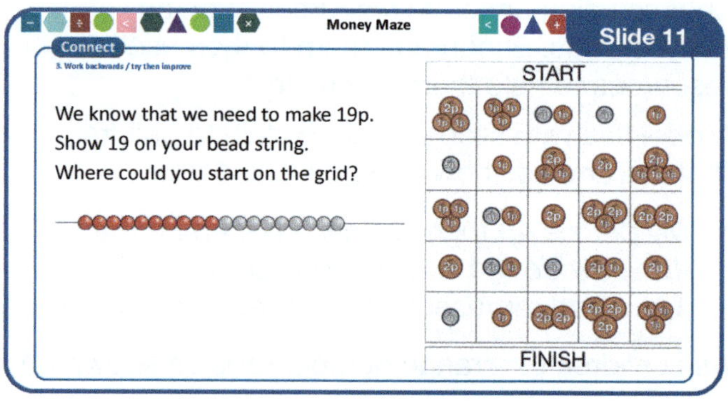

Display slide 11.

- *We know that we need to make 19p.*

Ask children to arrange their bead strings to show 19 beads.

- *What shall we do first?*

Children pick a starting point from the bottom of the grid and work backwards, moving beads to the right as they count back. The aim is to end up with no beads on the left-hand side.

Model the following path with children.

- *The 19 beads represent 19p.*
- *I am going to start at the 3p. I move 3 beads to the right. 19 – 3 is 16p.*
- *Next I will move to the 2p.*
- *I move 2 more beads to the right. 16 – 2 is 14p.*
- *Next I will move to the 4p. 14p – 4p = 10p.*
- *I know I have 10p left, so I need to look for combinations that make 10p.*
- *Next I will move to the 5p. So I now have 5p left.*
- *I will move to the 2p. I have 3p left.*
- *I can't move anywhere to take away 3p so this path does not work.*

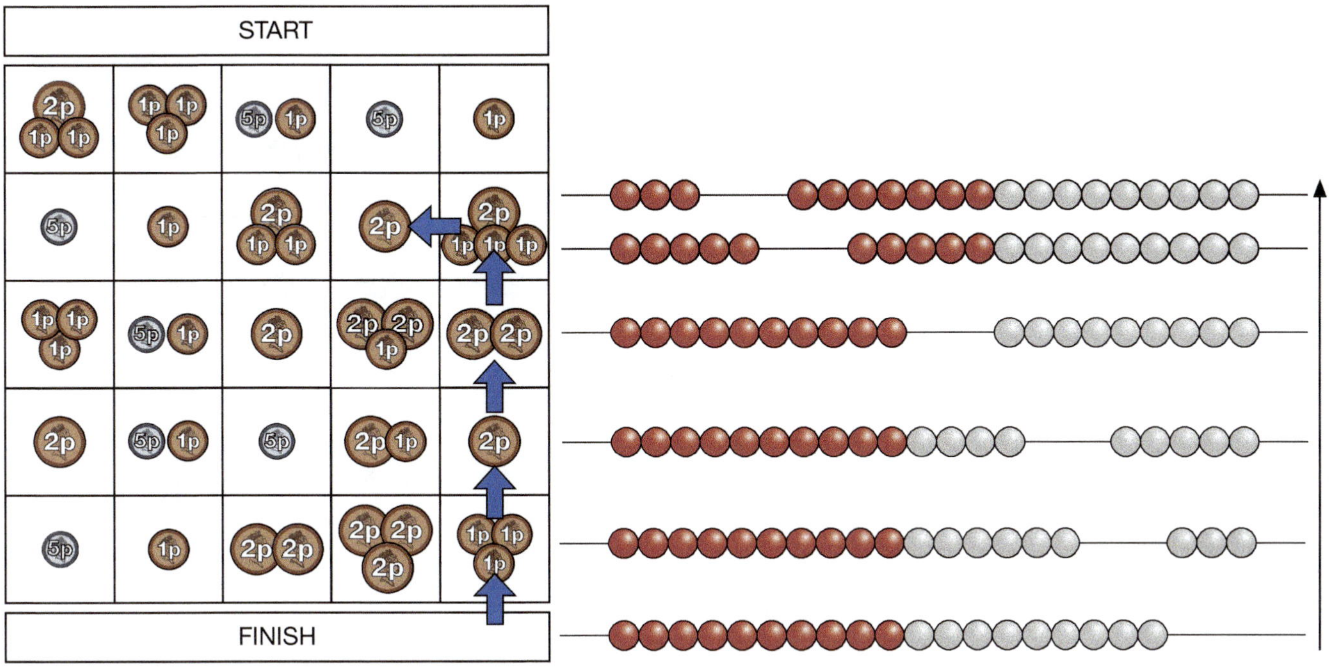

- *What could you change about your path which would mean that it makes exactly 19p? Do you need to start again?*

4. Create a systematic list

Children will need to consider how they will keep track of the paths that they have taken so that they do not repeat combinations.

Slide 12

Display slide 12.

- *How has this child kept track of their paths? What do you like / dislike about their work?*
- *What do you think they might do next? What would you do next?*

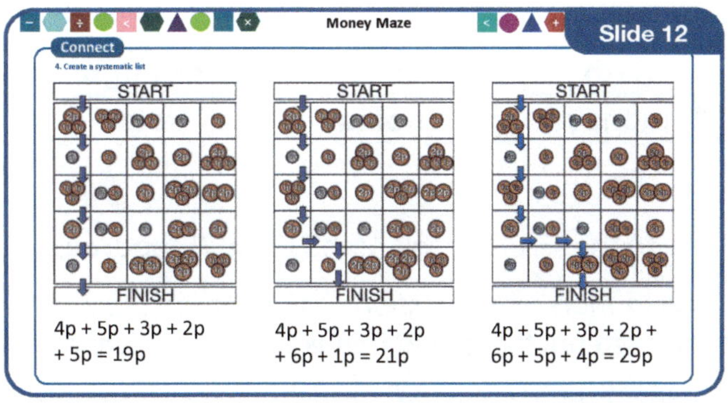

Slide 13

Print slide 13 and distribute to children. They could draw their paths on these grids, adapting them each time. They could then record the total below each grid.

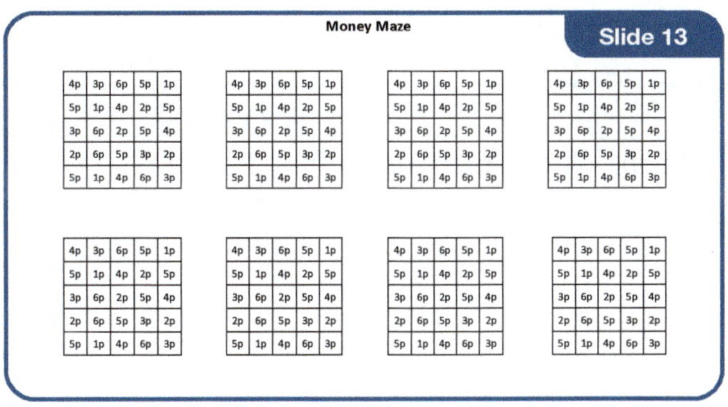

Collaborate

Refer to the **Collaborate** task of the **Pupil Resource**.

Slide 14

Display slide 14.

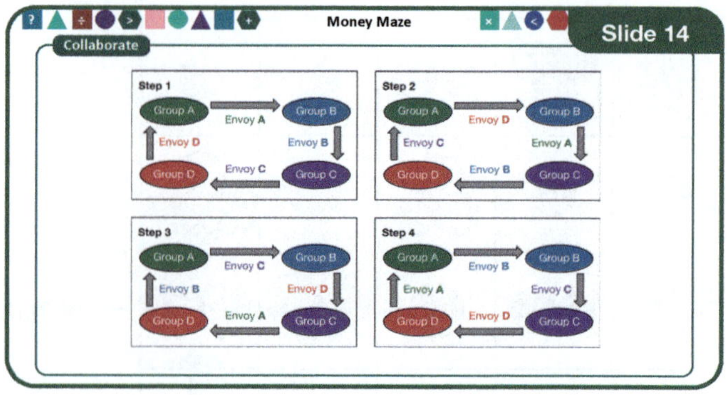

This task provides an opportunity for children to work independently, in pairs or in small groups to work through the maze.

As there are a number of solutions to this puzzle, this task provides a good opportunity for children to use the 'envoy' method. One child from each group is nominated as the envoy. Following discussion and exploration time, the envoy then moves to each of the other groups in turn, to ask questions about how they tackled the problem and the solutions they found, for example, the target total of 19p, highest value or lowest value path, before reporting back to their original group.

Communicate

Refer to the **Communicate** task of the **Pupil Resource**.

Slide 15

Display slide 15.

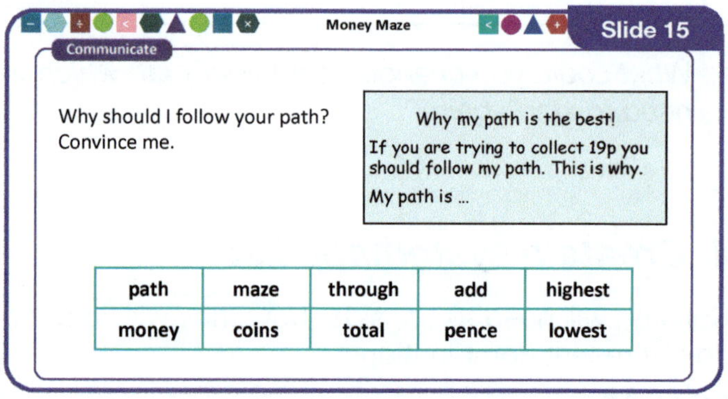

This task provides an opportunity for the children to create a short piece of persuasive writing. Alternatively, children could give a short speech about why their path is the best. They could use sound buttons or audio devices to record this.

Encourage them to use the vocabulary from the word bank.

path	**maze**	**through**	**add**	**highest**
money	**coins**	**total**	**pence**	**lowest**

Create

Refer to the **Create** task of the **Pupil Resource**.

- *Were there any questions asked at the start of the task that you now want to think about?*
- *Is there anything new that you are wondering now?*

Children can use the word bank to prompt further questions of their own.

across the board	**change**	**start**
finish	**fewest moves**	**highest total**

- <u>I wonder</u> what would happen if I changed the start and finish and went across the board instead of down. What would the highest total be?

Refer to the **Consider** task of the **Pupil Resource**.

Encourage children to think about the strategies that they have used. The prompts below may help get children started.

Cognitive learning domain

- The trickiest part of solving my question was …
- The new strategies I learned were …

Affective domain

- The fun bit of my learning was … It was fun because …
- It was easy / hard to keep track of the different paths because …
- I am / I am not pleased with my work because …

Metacognitive domain

- I used … to help me. It was helpful because …
- Next time, I will …
- A top tip I would give for working well as a group is …

Answers

There are many possible paths to 19p. For example:

4p + 5p + 3p + 2p + 5p = 19p

6p + 2p + 5p + 3p + 3p = 19p

1p + 5p + 4p + 2p + 3p + 4p = 19p

Solve a simpler but related problem

Four possible paths to 11p:

4p + 1p + 6p = 11p

6p + 3p + 2p = 11p

3p + 5p + 3p = 11p

5p + 2p + 4p = 11p

Grid Patterns

Year 2

Introduction

Grid Patterns provides children with an opportunity to explore patterns and rules when adding sets of consecutive numbers.

Children are presented with an addition grid, showing the totals of pairs of numbers from 1 to 10. Within the grid, 2 by 2 squares are highlighted, and children are asked to find the total of the four numbers. They will need to draw on a range of mental addition strategies, and start to explore multiplicative patterns. For example, 4 + 5 + 5 + 6 is equal to 4 × 5.

Efficiency of calculation is explored through use of concrete resources. Children use and compare a range of representations using the key question: What's the same? What's different?

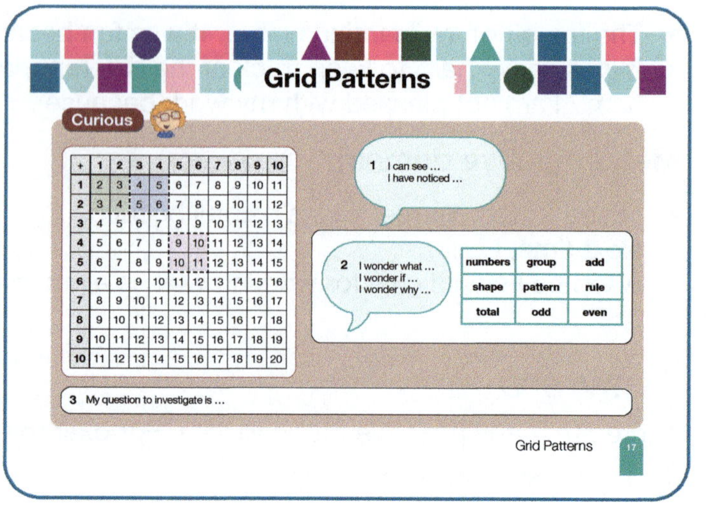

This task promotes greater-depth learning in a range of ways, supporting children to identify and articulate underlying mathematical structures. For example, in the **Create** stage, greater depth is promoted by supporting children to form their own questions that are creative and inventive.

This **Teacher Guide** is supported by photocopiable **Pupil Resources** and **Teacher's Presentation** PowerPoint slides for whole-class or group display.

National Curriculum

Grid Patterns covers these Year 2 National Curriculum objectives.

Addition and subtraction

- Solve problems with addition and subtraction:
  - using concrete objects and pictorial representations, including those involving numbers, quantities and measures
  - applying their increasing knowledge of mental and written methods.
- Recall and use addition and subtraction facts to 20 fluently, and derive and use related facts up to 100.

The 6C cycle

The grid on the right shows the relative weighting given to each of the 6 Cs in **Grid Patterns**.

Curious	★★
Connect	★★★
Collaborate	★★
Communicate	★★
Create	★★★
Consider	★

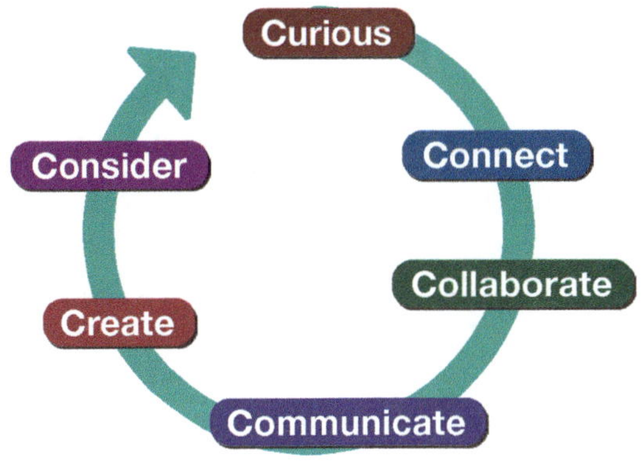

Problem-solving heuristics

The grid shows which problem-solving strategies children are likely to use when investigating the Connect tasks in **Grid Patterns.** These strategies are further exemplified in the Connect stage below (pages 63–66).

Draw a diagram	✓
Make a table / spot a pattern	✓
Make a model / act it out	✓
Try then improve	
Create a systematic list	
Solve a simpler but related problem	
Work backwards	

Key mathematical ideas

Children may choose from a range of strategies to calculate the total of the numbers in a square.

For example, for this 2 by 2 square, the following strategies may be used:

2	**3**
3	**4**

- bonds to 6 (2 + 4 + 3 + 3).
- bonds to 10 (3 + 3 + 4 + 2).
- doubles (double 3 + 4 + 2).
- rebalancing (2 + 4 has equal value to 3 + 3 so 3 + 3 + 3 + 3 could be used).

Children may begin to notice the relationship with multiplication. If using the rebalancing strategy, they may be able to work out 4 × 3 (4 groups of 3) by doubling 3 twice.

For other sets of numbers, they may be able to recall a multiplication fact. Rebalancing the four numbers in this square, for example, would allow them to calculate the total using 4 × 10 = 40, which may be a known fact.

9	**10**
10	**11**

If children notice a pattern or rule and then choose to explore the larger numbers on the grid, they may use more multiplicative strategies.

For example:

16	**17**
17	**18**

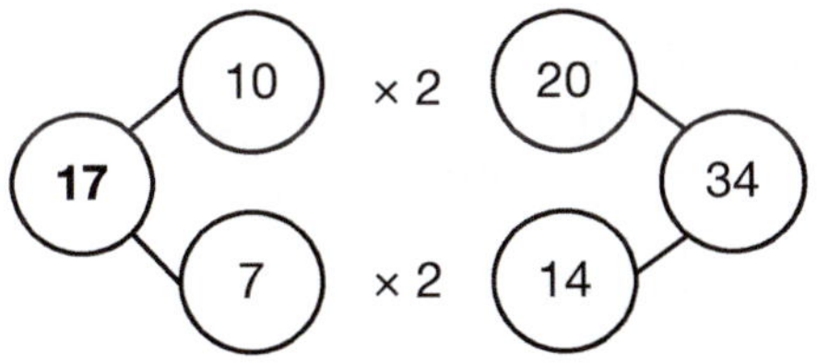

Then: double 30 = 60 double 4 = 8 total = 68

If children have not used a rebalancing strategy and noticed the link to doubles, they may calculate double 17 and then use a more formal written strategy to add 16 and 18 before combining to find a total.

Answers are given in situ throughout the tasks because of the cumulative nature of the tasks.

Teaching guidance

If you are using the **Express Pathway**, please turn to pages 63–68.

Teaching guidance for the **Full 6C Pathway** is given below.

The **Teacher's Presentation** PowerPoint file has slides specifically mentioned in this teaching guidance. Some of the slides include notes (within the PowerPoint file) providing 'stumble support' suggestions, that may help you to support children in their problem-solving thinking and enable access for children who are less confident.

Curious

Awe and wonder

Slide 1

Display slide 1.

Ask children to share what they are noticing or wondering.

- I can see a grid with numbers.
- I can see coloured squares.
- I have noticed that there are four numbers in each square.
- I have noticed that each square has a double number.
- I can see three squares.

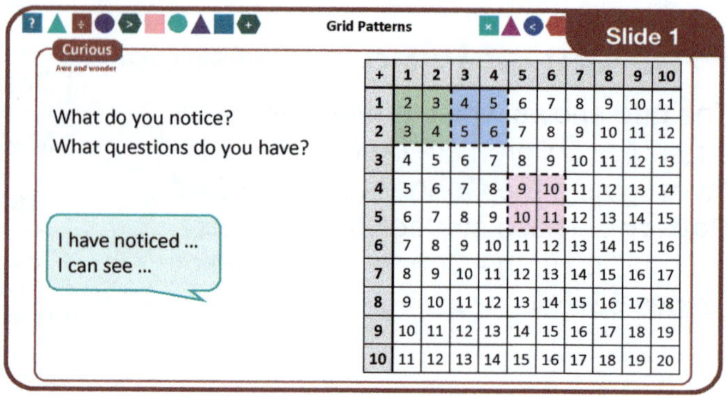

+	1	2	3	4	5	6	7	8	9	10
1	2	3	4	5	6	7	8	9	10	11
2	3	4	5	6	7	8	9	10	11	12
3	4	5	6	7	8	9	10	11	12	13
4	5	6	7	8	9	10	11	12	13	14
5	6	7	8	9	10	11	12	13	14	15
6	7	8	9	10	11	12	13	14	15	16
7	8	9	10	11	12	13	14	15	16	17
8	9	10	11	12	13	14	15	16	17	18
9	10	11	12	13	14	15	16	17	18	19
10	11	12	13	14	15	16	17	18	19	20

Develop and deepen

Encourage children to use the words in the word bank to develop and deepen their thoughts and questions.

numbers	**group**	**add**
shape	**pattern**	**rule**
total	**odd**	**even**

Possible lines of enquiry:

- I wonder what each group's total is.

Possible lines of enquiry linked to the word bank:

- I wonder what I would notice if I added the four numbers in each group.
- I wonder if there is a pattern in the numbers.
- I wonder if there is a rule for finding the group totals.

Possible lines of enquiry that children may raise now, but should be held back until the **Create** task:

- I wonder why all of the totals are even.
- I wonder if there is a link between totals if groups overlap.
- I wonder what would happen if we added a group of numbers in a bigger square.
- I wonder what would happen if we changed the shape of each group.

Consider whether / how to collate and display all the suggested questions, and whether / how to steer children towards selecting an appropriately challenging question.

Pupil Resources for Full and Express Pathways

Distribute the **Pupil Resources** for **Grid Patterns** to the children to help support them in their exploration of the different tasks. Use either of the **Curious Pupil Resources** depending on whether you are following the **Full Pathway** or the **Express Pathway**. The **Pupil Resources** for the remaining 5Cs are common to both pathways.

To provide additional flexibility in the classroom, each of the **Pupil Resources** is also provided in the **Teacher's Presentation** PowerPoint file (see slides 2–5), so you can display these at any time if you wish.

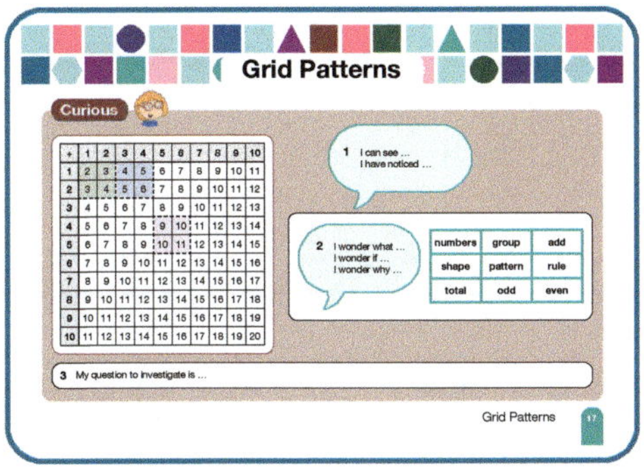

or

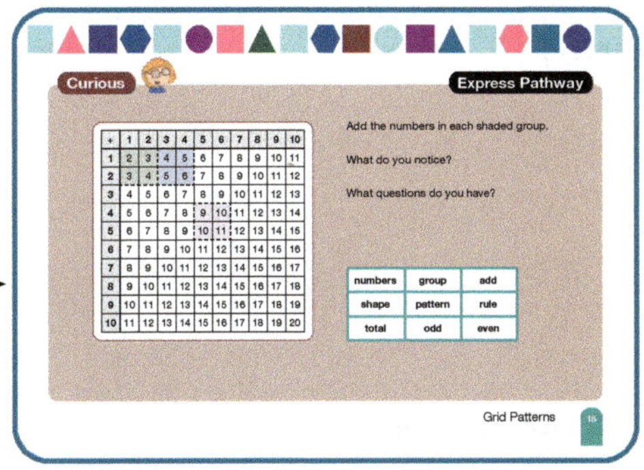

then

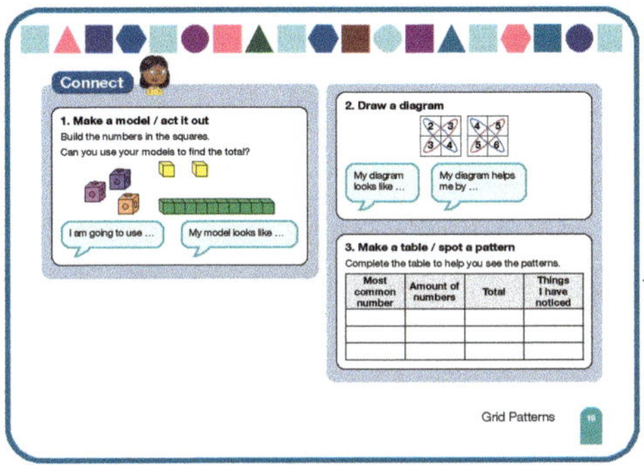

and

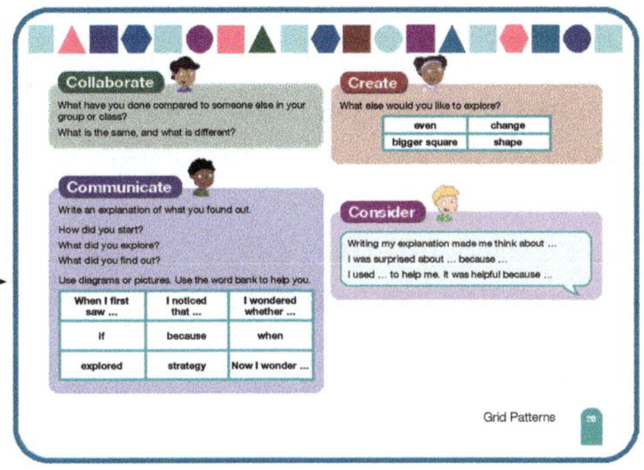

Express Pathway

If you are following the **Express Pathway**, spend a few minutes discussing the **Curious Pupil Resource** with the children. (This is slide 3, so you could display it if you wish.) Use some of the prompts on the previous page to help.

Three tasks involving different strategies are suggested for children to make connections.

Slide 6

Print slide 6 and distribute to children. They can use this to investigate the squares highlighted in the **Curious** task, but it also provides them with the opportunity to choose their own 2 by 2 squares, or larger squares, to explore.

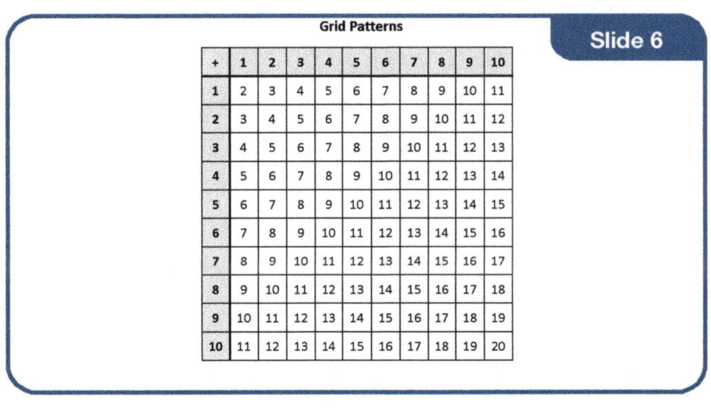

+	1	2	3	4	5	6	7	8	9	10
1	2	3	4	5	6	7	8	9	10	11
2	3	4	5	6	7	8	9	10	11	12
3	4	5	6	7	8	9	10	11	12	13
4	5	6	7	8	9	10	11	12	13	14
5	6	7	8	9	10	11	12	13	14	15
6	7	8	9	10	11	12	13	14	15	16
7	8	9	10	11	12	13	14	15	16	17
8	9	10	11	12	13	14	15	16	17	18
9	10	11	12	13	14	15	16	17	18	19
10	11	12	13	14	15	16	17	18	19	20

1. *Make a model / act it out*

Additional resources required: interlocking cubes, base-10 blocks

To explore the relationship between the numbers within a 2 by 2 square, and the square's total value, children could represent the set of numbers with interlocking cubes.

Slide 7

Display slide 7.

- *Show the numbers in this square using cubes. Can you use a different colour for each number?*

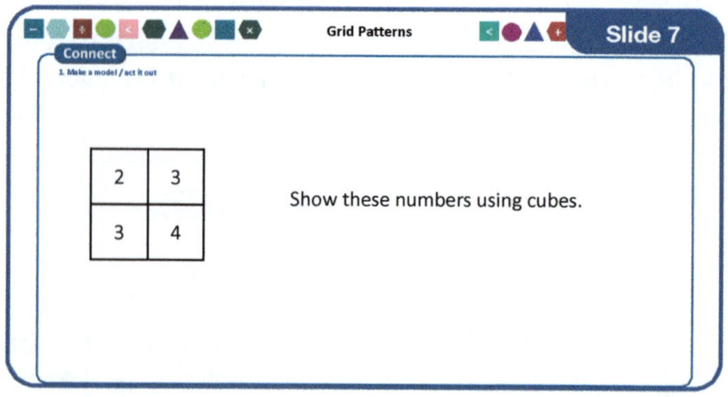

Slide 8

Display slide 8. Ask children to compare their towers with those shown on the slide.

- *What do you notice about the towers?*
- Two of the towers have the same number of cubes, one has 1 less, and the other has 1 more.
- *What could we do with the cubes to make adding the four numbers easier?*

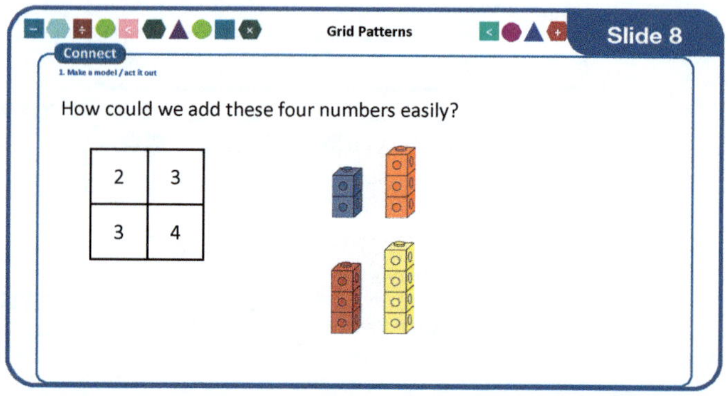

Slide 9

Display slide 9.

Support children to understand the rebalancing strategy shown by the rearrangement of the cubes. By moving one from the largest number to the smallest number, the sets have equal value.

- *Could you use this strategy with other number squares?*

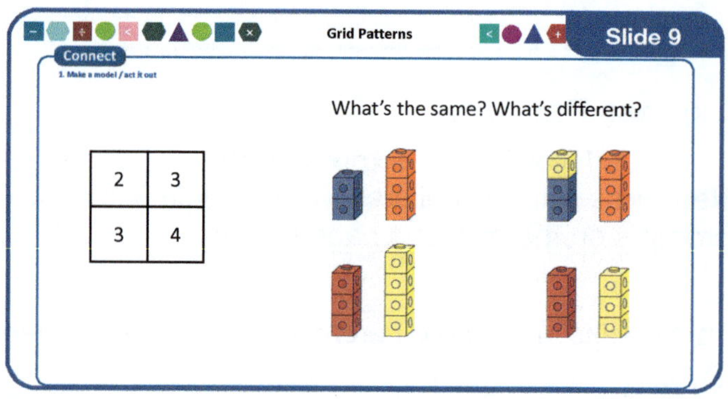

Give children time to explore on their blank grids (slide 6).

When children explore larger numbers, you may wish to suggest that they use base-ten apparatus for more efficient representation. For example:

16	**17**
17	**18**

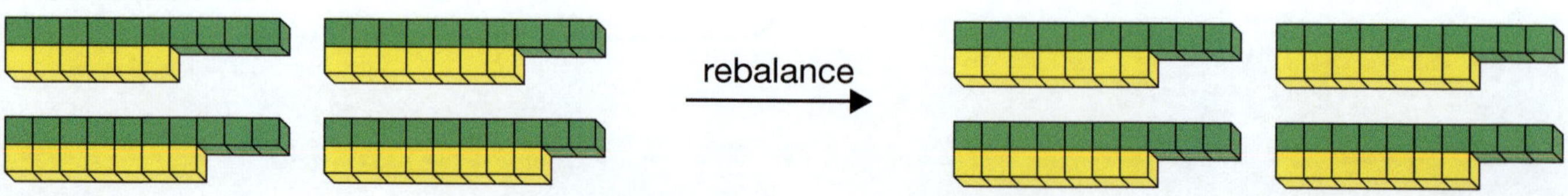

2. Draw a diagram

Slide 10

Display slide 10.

- *What do you notice about the sets of numbers that are circled?*

Support children to recognise that in each square there are two numbers that are the same, one number that is 1 less and one that is 1 more. Ask them to think about how they can use this knowledge to find the total of the four numbers.

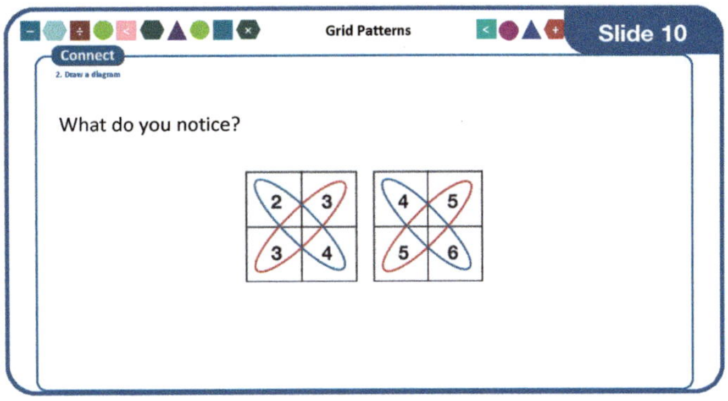

- *How can you make adding the numbers easier?*
- *Can you make 4 lots of 3 for the first square and 4 lots of 5 for the second?*

After they have explored the two squares on the slide, ask children to circle numbers on their grids to explore other sets of four numbers.

Some children may wish to explore how this strategy works with a larger set of numbers, and then pursue this further in the **Create** task.

Slide 11

Display slide 11, which shows a 3 by 3 square with 10 in the centre square.

- *What do you notice?*
- *How could you find 9 tens?*

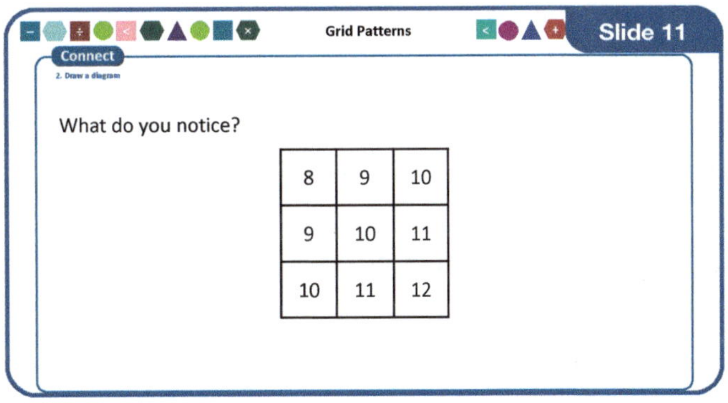

Ask children to draw circles around pairs of numbers that could be rebalanced to make tens.

You may wish to model this on the slide. For example:

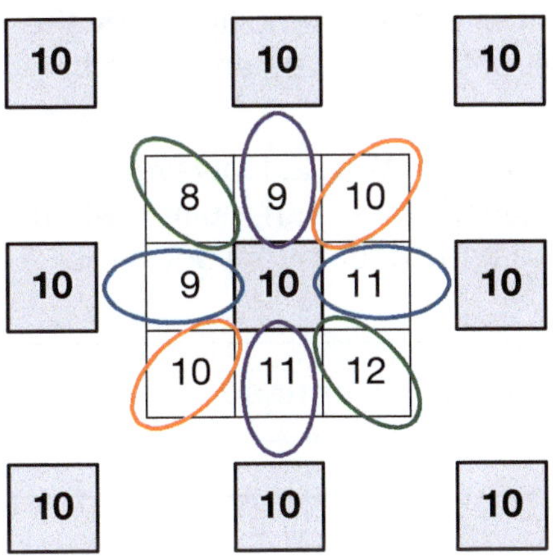

The 9 lots of 10 come from:

- three pairs of numbers (8 and 12, 9 and 11, 9 and 11) that could be rebalanced to each make 2 tens;
- 2 tens in the bottom-left and top-right squares;
- one 10 in the centre of the large square.

Using the model above, children could explore other sets of 9 numbers.

Connect

3. Make a table / spot a pattern

Slide 12

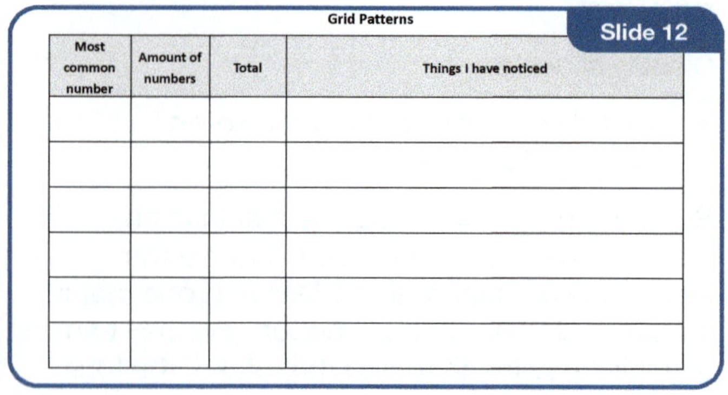

Print slide 12 and distribute to children.

To further explore the link between the total additive value of the squares and multiplication, children could make a table.

Slide 13

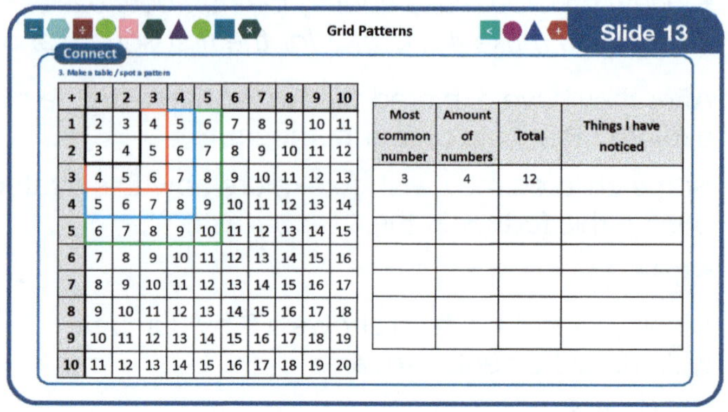

Display slide 13 and discuss the growing square size.

- *How many numbers does the first square have?* (4)
- *Which number is repeated in that square?* (3)
- *What total do you get when you add up those 4 numbers?* (12)
- *What about the next one? How many numbers are there?*
- There are 9 numbers and the most common number is 4.

Model filling in the first couple of rows in the table.

- *What do you notice?*
- *Can you see a link?*

What strategy could you use to quickly add up the total of your square?

Support children where necessary to see that the most common number in the set multiplied by the amount of numbers in the set is equal to the total value of the set if all of the numbers are added together.

Some of these multiplication tables are not within the National Curriculum for Year 2. However, children can solve them using repeated addition. Once they have identified a link with multiplication, children could use this as an opportunity to find creative ways of using known facts to work out unknown facts.

Most common number	Amount of numbers	Total	Things I have noticed
3	4	12	3 + 3 + 3 + 3
4	9	36	1 group of 4 fewer than 10 groups of 4
5	16	80	10 × 5 = 50 6 × 5 = 30
6	25	150	4 lots of 25 = 100 2 lots of 25 = 50

Example of slide 12 table

Refer to the **Collaborate** task of the **Pupil Resource**.

This task provides an opportunity for children to work together to calculate the totals from different sets of numbers in the main grid and then to come up with a rule together.

Ask children to work using their chosen strategy and then share this with the group. Other members should then explore whether a different strategy gets them to the same answer.

- *What's the same?*
- *What's different?*

Collaborate

Refer to the **Communicate** task of the **Pupil Resource**.

Slide 14

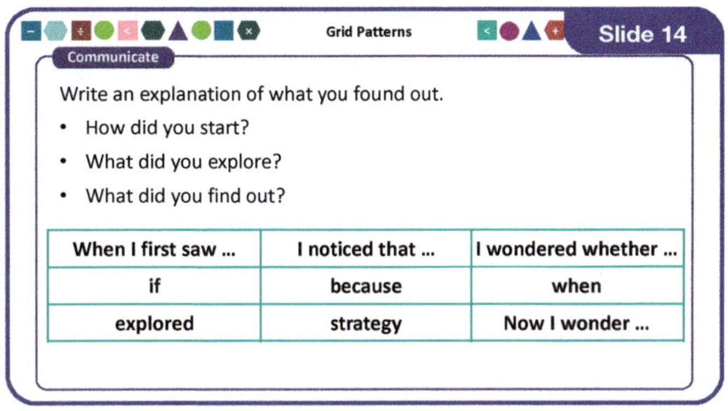

Display slide 14.

This task provides an opportunity for the children to write an explanation of how they chose the numbers to explore and how they found each total.

This could include subheadings to define each part of the task. Ask them to detail where they started, what they explored and what they found out. Encourage them to use their diagrams or models to illustrate their thought process during the **Connect** stage. A word bank is provided to support their use of vocabulary.

When I first saw ...	**I noticed that ...**	**I wondered whether ...**
if	**because**	**when**
explored	**strategy**	**Now I wonder ...**

They may choose to include some things they are left wondering from the **Create** task and also some of their reflections from the **Consider** task.

Communicate

Create

Refer to the Create task of the **Pupil Resource**.

Slide 15

Display slide 15.

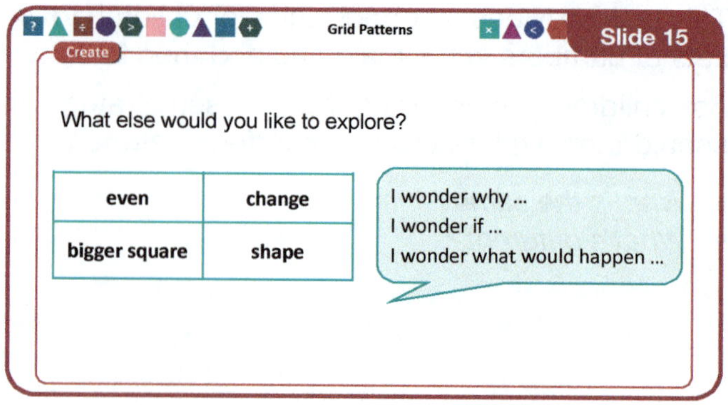

- *Were there any questions asked at the start of the task that you now want to think about?*
- *Is there anything new that you are wondering now?*

Children can use the word bank to prompt further questions of their own.

even	**change**
bigger square	**shape**

- I wonder why all of the totals are even.
- I wonder if there is a link between totals if groups overlap.
- I wonder what would happen if we added a group of numbers in a bigger square.
- I wonder what would happen if we changed the shape of each group.

The totals are always even because there is always a double within the addition and when the remaining numbers are rebalanced, another double is created.

Consider

Refer to the Consider task of the **Pupil Resource**.

Encourage children to think about the strategies that they have used. The prompts below may help get children started.

Cognitive learning domain

- A new strategy I learned was …
- Writing my explanation made me think about …

Affective domain

- I was surprised about … because …
- I am pleased / not pleased with my work because …

Metacognitive domain

- I got stuck when … I unstuck myself by …
- I used … to help me. It was helpful because …
- Next time, I will …

All the Arrays

Year 1

Introduction

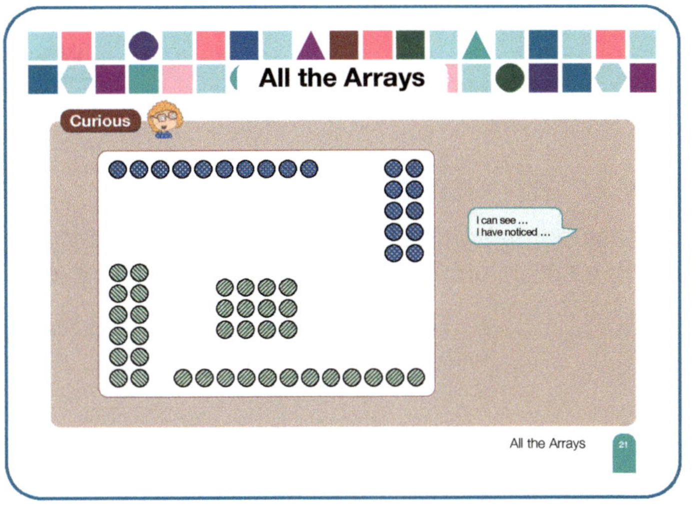

All the Arrays allows children to explore deeply the concept of arrays, which lays the foundations for multiplicative understanding across the primary phase and beyond. The task centres on exploring the idea that some numbers can be arranged in many arrays (for example, 8 × 1, 1 × 8, 2 × 4, 4 × 2), while some numbers can only be arranged into two arrays (for example, 7 × 1 and 1 × 7). In Upper Key Stage 2 children will use the language of multiples, factors, square numbers, prime numbers and composite numbers to describe and apply this knowledge. For this task, the focus is firmly on the possible arrays that can be built.

This task promotes greater-depth learning in a range of ways, supporting children to identify and articulate underlying mathematical structures. For example, in the **Communicate** stage, greater depth is promoted by supporting children to communicate their thoughts clearly, accurately and in a well-organised way.

This **Teacher Guide** is supported by photocopiable **Pupil Resources** and **Teacher's Presentation** PowerPoint slides for whole-class or group display.

National Curriculum

All the Arrays covers these Year 1 National Curriculum objectives.

Multiplication and division

- Solve one-step problems involving multiplication and division, by calculating the answer using concrete objects, pictorial representations and arrays with the support of the teacher.

The 6C cycle

The grid on the right shows the relative weighting given to each of the 6 Cs in **All the Arrays**.

Curious	★
Connect	★★
Collaborate	★
Communicate	★★★
Create	★★
Consider	★★

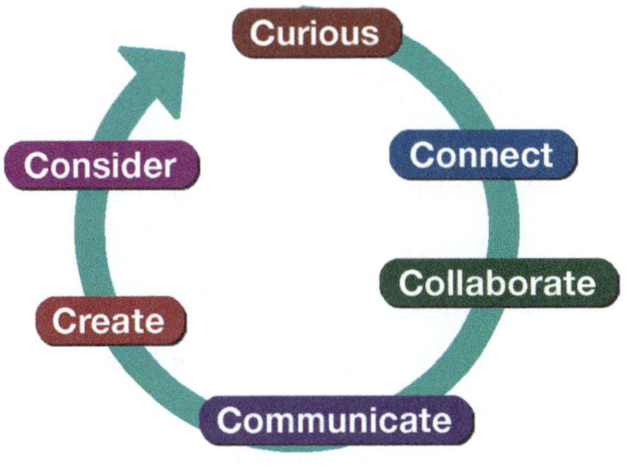

Problem-solving heuristics

The grid shows which problem-solving strategies children are likely to use when investigating the **Connect** tasks in **All the Arrays**. These strategies are further exemplified in the **Connect** stage below (pages 73–75).

Draw a diagram	✓
Make a table / spot a pattern	✓
Make a model / act it out	✓
Try then improve	
Create a systematic list	✓
Solve a simpler but related problem	
Work backwards	

Key mathematical ideas

This task centres on the concept of arranging dots into a rectangular array with an equal number of dots in each column and in each row. For example:

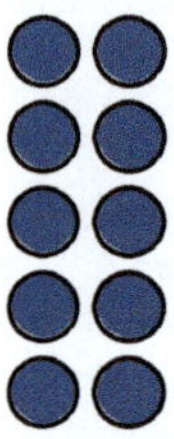

- This array has a total of 10 dots.
- I can see 2 + 2 + 2 + 2 + 2.
- I can see 5 + 5.
- I can see 2 groups of 5.
- I can see 5 groups of 2.
- There are 2 dots in each row. There are 5 dots in each column.

Although there are clear links to multiplication, children do not need to know multiplication facts for this task. It can be explored simply through moving the dots into different arrangements and working systematically to find all the possible arrays. Children will discover that the size of the number does not always link to the number of possible arrays.

There are likely to be discussions about the difference between, for example, a 3 × 4 array and a 4 × 3 array. Do children recognise that they are different, but very closely linked, with the same total?

In this task, arrays are linked to multiplication facts in the following way:

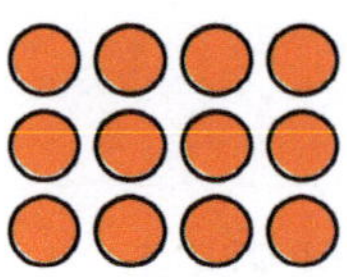

3 rows, 4 columns
= 3 groups of 4
= 3 × 4

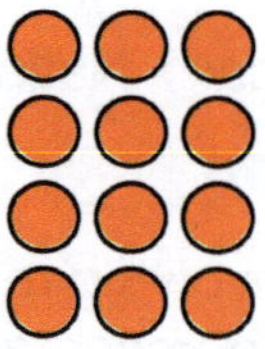

4 rows, 3 columns
= 4 groups of 3
= 4 × 3

⚠ Do all the children understand the concept of an array? If not, explain that an array is a collection of dots, counters or other objects ordered into rows and columns, with an equal number of items in each row / column. If needed, allow them some time to explore arranging counters into formations with equal and unequal rows and columns.

Answers to the tasks are given on page 77.

Teaching guidance

Teaching guidance is given below.

The **Teacher's Presentation** PowerPoint file has slides specifically mentioned in this teaching guidance. Some of the slides include notes (within the PowerPoint file) providing 'stumble support' suggestions, that may help you to support children in their problem-solving thinking and enable access for children who are less confident.

Awe and wonder

Slide 1

Display slide 1.

Ask children to share what they are noticing or wondering.

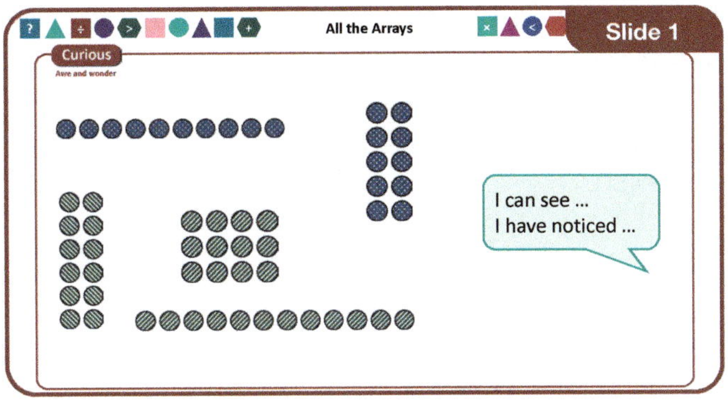

- I can see some blue spotty dots and some green striped dots.
- I have noticed that the two patterns with blue spotty dots have the same total number of dots: 10.
- I have noticed that there are two sets of blue spotty dots but three sets of green striped dots.
- I can see all the dots are in the shape of different rectangles.

Develop and deepen

Slide 2

Display slide 2.

Encourage children to use the words in the word bank to develop and deepen their thoughts and questions.

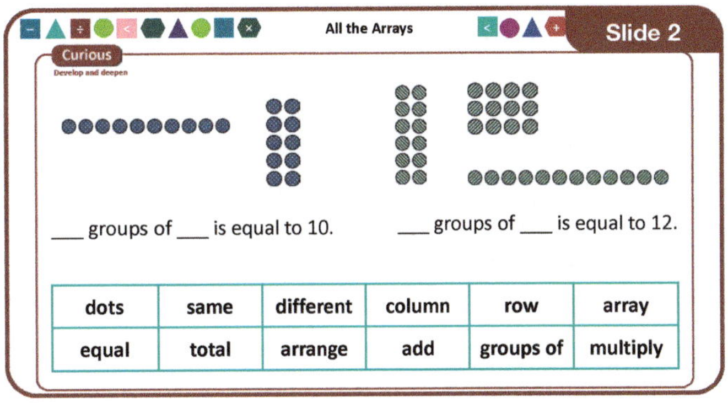

dots	**same**	**different**
column	**row**	**array**
equal	**total**	**arrange**
add	**groups of**	**multiply**

Possible lines of enquiry linked to the word bank:

- I wonder if the blue spotty dots can be arranged in a different way.
- I wonder why the dots have been arranged like rectangles.
- I wonder why each set has the same number of dots in each row and each column.
- I wonder how many different ways there are to arrange the dots into a rectangle.
- I wonder if bigger numbers have more ways of creating different arrays.

Possible lines of enquiry that children may raise now, but should be held back until the **Create** task:

- I wonder if every number can be arranged into equal groups of 2 / 3 / 4.

Consider whether / how to collate and display all the suggested questions, and whether / how to steer children towards selecting an appropriately challenging question.

- *Do you think these are all the possible arrays for 10?*

Model creating a 2 by 5 array, and a 5 by 2 array, and show them side by side.

- *Are these the same or different?*

Agree that they are very similar, but children should count them as two different arrays.

Pupil Resources

Distribute the **Pupil Resources** for **All the Arrays** to the children to help support them in their exploration of the different tasks.

To provide additional flexibility in the classroom, each of the **Pupil Resources** is also provided in the **Teacher's Presentation** PowerPoint file (see slides 3–6), so you can display these at any time if you wish.

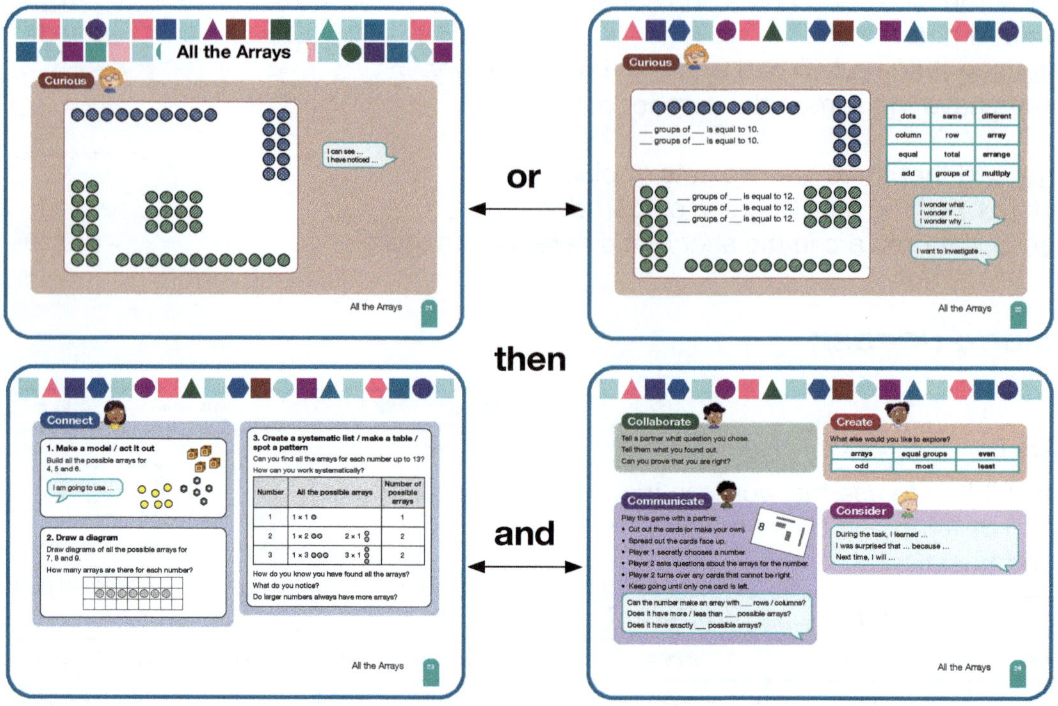

Three tasks involving different strategies are suggested for children to make connections.

1. Make a model / act it out

Additional resources required: concrete resources that can be arranged into arrays (for example, counters, cubes, peg boards and pegs, beads, 2D shapes), camera (optional)

Slide 7

Display slide 7.

- *How many arrays can you make using four objects?*

Encourage children to make a concrete model of the dots.

- *How do you know when you have found an array?*
- The rows / columns all have the same number of objects.

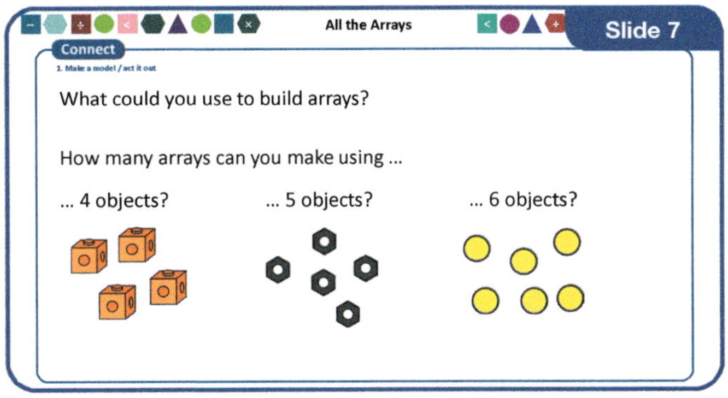

Encourage children to keep each model they make, so they create a collection of arrays.

- *How are you going to make sure you have found all the possible arrays for 4?*

When children have found all the possible arrays, you could take a photograph of their models.

- *Do you think there will be more possible arrays for 5 than for 4? Why / why not?*
- *Build all the arrays for 5.*

Then ask children to repeat for 6.

2. Draw a diagram

Additional resources required: concrete resources that can be arranged into arrays (optional), squared paper (optional)

- *How many arrays can you make using 7 dots?*

Some children may want to build the arrays using concrete resources. Others may want to go straight to drawing diagrams.

- *Why do you think it could be helpful to keep a record of the arrays you have found?*
- Because we can show people what we've done. If we only build arrays using cubes, the cubes will have to be put away and then the models are gone.

Slide 8

Print slide 8 and distribute to the children. Ask children to record the arrays for 7 by drawing dots in the section of squared paper. They can then record the total number of arrays in the final column of the table.

- *Now I'd like you to do the same for 8, and then 9.*

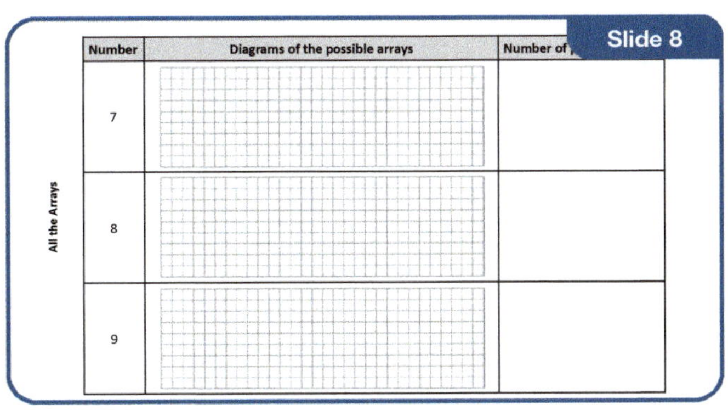

Number	Diagrams of the possible arrays	Number of possible arrays
7	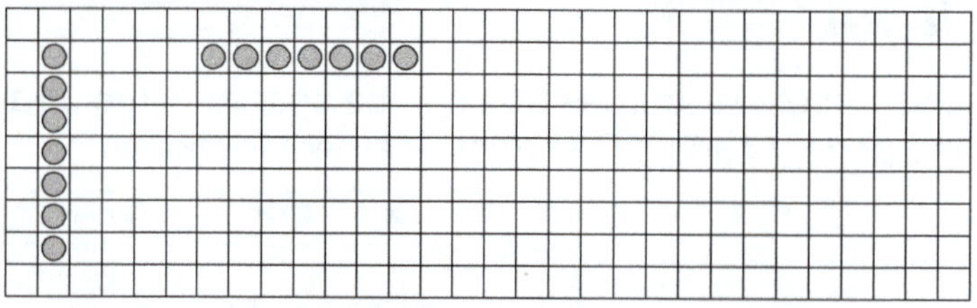	2
8	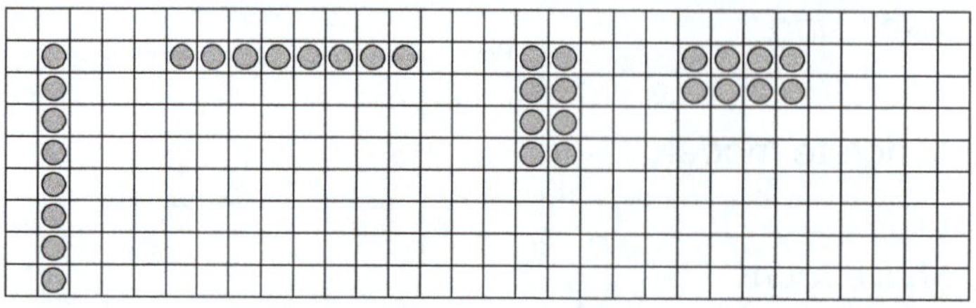	4
9	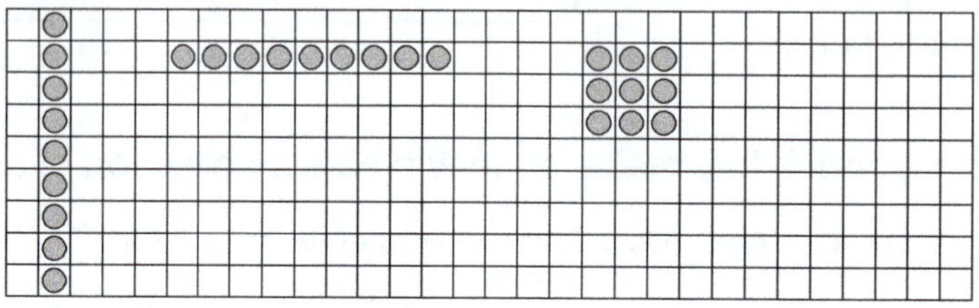	3

Example of a completed table

If children would like to repeat this for some other numbers, provide them with sheets of squared paper to continue their recording.

- *Look at the numbers of possible arrays for each number. Are you surprised? Why / why not?*

3. Create a systematic list / make a table / spot a pattern

Slides 9 and 10

Print slides 9 and / or 10 and distribute to children as needed. Slide 10 provides a greater level of support.

Some children may want to create a systematic list, organised into a table. They may also want to make models or draw diagrams to support them in this approach.

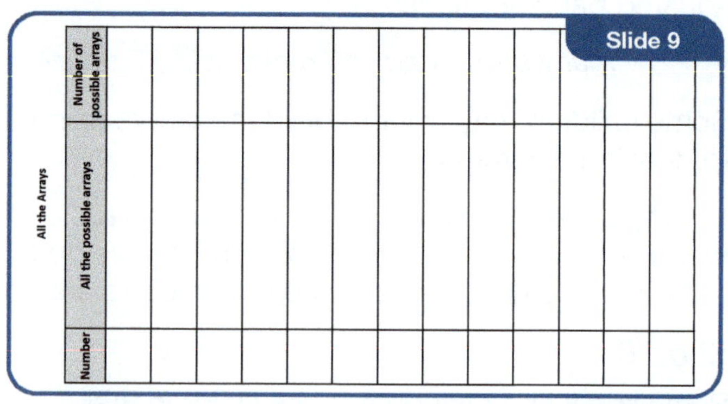

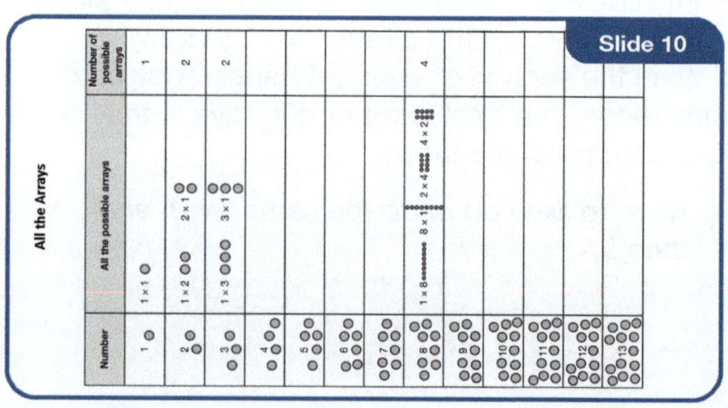

Slide 11

Display slide 11.

- *Can you see how the table on the slide helps to keep everything organised?*
- *What do you notice about the numbers in the left-hand column?*
- They are in order, starting from 1.

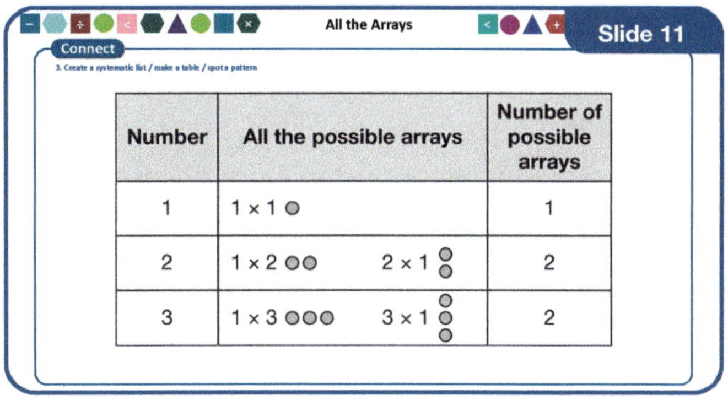

Encourage children to use their print-out of either slide 9 or slide 10 to work through the numbers in a systematic order, recording all the possible arrays and the number of arrays possible.

Number	All the possible arrays	Number of possible arrays
1	1 × 1	1
2	1 × 2 2 × 1	2
3	1 × 3 3 × 1	2
4	1 × 4 4 × 1 2 × 2	3
5	1 × 5 5 × 1	2
6	1 × 6 6 × 1 2 × 3 3 × 2	4
7	1 × 7 7 × 1	2
8	1 × 8 8 × 1 2 × 4 4 × 2	4
9	1 × 9 9 × 1 3 × 3	3
10	1 × 10 10 × 1 2 × 5 5 × 2	4
11	1 × 11 11 × 1	2
12	12 × 1 12 × 1 2 × 6 6 × 2 3 × 4 4 × 3	6
13	1 × 13 13 × 1	2

Example of a completed table

- *How do you know that you have found all the possible arrays for each number?*
- *What do you notice about the numbers of possible arrays?*

Refer to the **Collaborate** task of the **Pupil Resource**.

Ask children to work in pairs or small groups. Ask them to each say what question they chose to investigate at the start of the task and explain what they have found out.

- *What was the question you thought of at the start of the task?*
- *Do you think you know the answer to that question now?*
- *How can you explain what you have found out? Can you convince your partner that you are right?*

This task also provides an opportunity for children to collaborate with a partner to play a game. See the **Communicate** task for more details.

Communicate

Refer to the **Communicate** task of the **Pupil Resource**.

Slide 12

Display slide 12.

This task provides an opportunity for children to apply and communicate what they know to a card game based on numbers and their possible arrays.

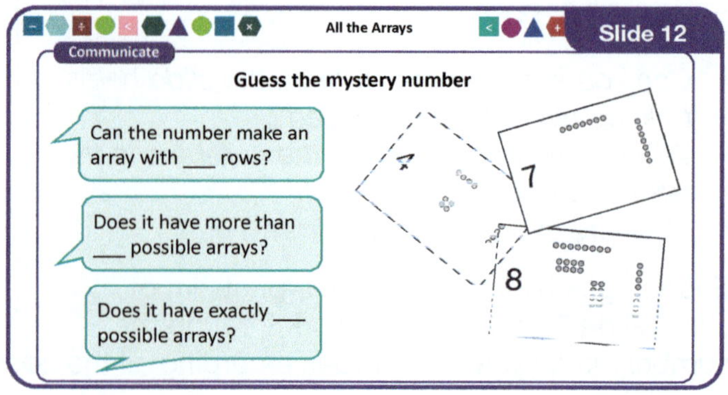

Slide 13

If you would like children to play the game suggested here, you could print slide 13, distribute to the children and ask them to cut out the cards. Alternatively, children could make their own cards, and either play the same game, or their own variation of it.

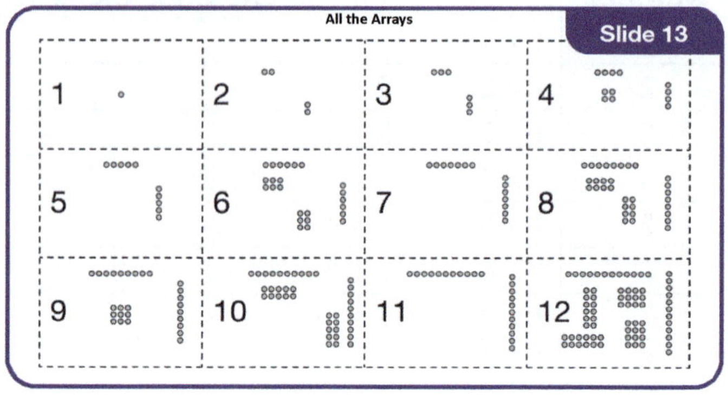

How to play:

Children spread out the cards face up.

Player 1 secretly chooses a card (without picking it up).

Player 2 asks 'yes' or 'no' questions. After each question, Player 2 turns over any cards that do not match the information. When only one card remains, Player 1 confirms that the correct number has been identified.

For example:

- *Can the number make an array with rows of 3?* (Yes. Player 2 turns over 1, 2, 4, 5, 7, 8, 10 and 11.)

Create

Refer to the **Create** task of the **Pupil Resource**.

Slide 14

Display slide 14.

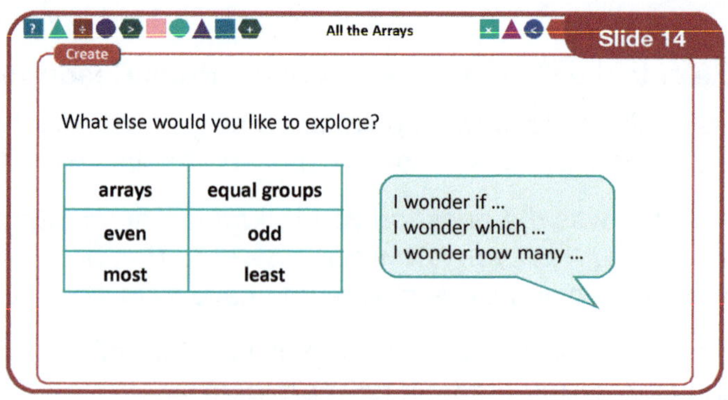

- *Were there any questions asked at the start of the task that you now want to think about?*
- *Is there anything new that you are wondering now?*

Children can use the word bank to prompt further questions of their own.

arrays	**equal groups**	**even**
odd	**most**	**least**

- I wonder if every number can be arranged into equal groups of 2 / 3 / 4.
- I wonder which number has the most possible arrays.
- I wonder which numbers have an odd number of possible arrays.
- I wonder how many numbers only have two possible arrays.
- I wonder if I can find arrays in real life.

Refer to the **Consider** task of the **Pupil Resource**.

Slide 15

Display slide 15.

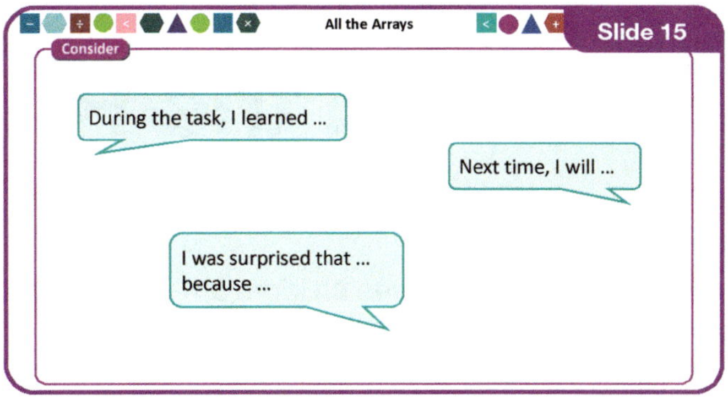

Encourage children to think about the strategies that they have used. The prompts below may help get children started.

Cognitive learning domain

- The trickiest part of solving my question was …
- During the task, I learned …

Affective domain

- … was fun because …
- I said 'aha' when … because …
- I was surprised that … because …

Metacognitive domain

- I got stuck when … I unstuck myself by …
- I used … to help me. It was helpful because …
- Next time, I will …

Answers

There is no expectation that children will work with numbers beyond 13, but some may choose to explore larger numbers, perhaps during the **Create** task. Here are the number of possible arrays for numbers 1–30:

Number	Number of possible arrays
1	1
2	2
3	2
4	3
5	2
6	4
7	2
8	4
9	3
10	4

Number	Number of possible arrays
11	2
12	6
13	2
14	4
15	4
16	5
17	2
18	6
19	2
20	6

Number	Number of possible arrays
21	4
22	4
23	2
24	8
25	3
26	4
27	4
28	6
29	2
30	8

Crossing the River

Year 1 Year 2

Introduction

Crossing the River gives children an opportunity to practise counting in steps of 2 in the context of Freddie the frog landing on every other lily pad to reach the river bank. Children will discover patterns with steps of 2, and this may lead them to generalise about the number of jumps needed for any number of lily pads.

For Year 2 children, the task could be extended to include steps of 3 by allowing Freddie to land on every third lily pad.

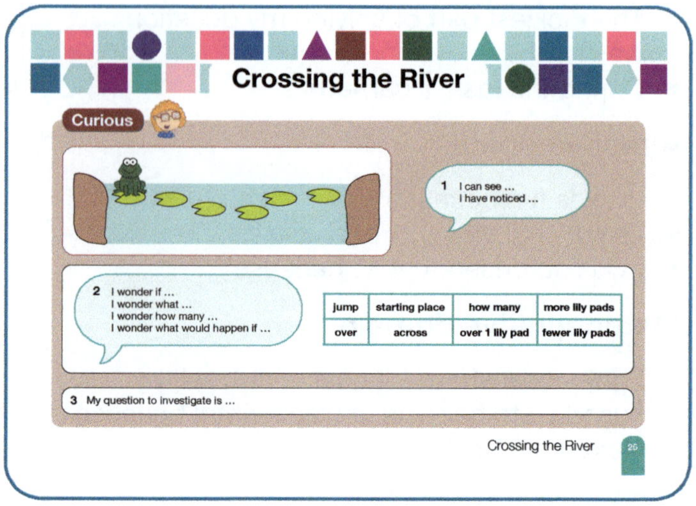

Children will act out the scenarios, or build models using concrete resources, then present their results in a structured way to enable pattern spotting. They will then create diary entries in the role of Freddie to explore a further way of showing what they have discovered.

This task promotes greater-depth learning in a range of ways, supporting children to identify and articulate underlying mathematical structures. For example, in the **Connect** stage, greater depth is promoted by supporting children to choose an appropriate strategy and explain their choice, and understand that sometimes more than one answer is possible.

This **Teacher Guide** is supported by photocopiable **Pupil Resources** and **Teacher's Presentation** PowerPoint slides for whole-class or group display.

National Curriculum

Crossing the River covers these Year 1 and Year 2 National Curriculum objectives.

Year 1 Number and place value

- Count, read and write numbers to 100 in numerals; count in multiples of twos, fives and tens.

Year 1 Multiplication and division

- Solve one-step problems involving multiplication and division, by calculating the answer using concrete objects, pictorial representations and arrays with the support of the teacher.

Year 2 Number and place value

- Count in steps of 2, 3, and 5 from 0, and in tens from any number, forward and backward.

Year 2 Multiplication and division

- Solve problems involving multiplication and division, using materials, arrays, repeated addition, mental methods, and multiplication and division facts, including problems in contexts.

The 6C cycle

The grid on the right shows the relative weighting given to each of the 6 Cs in **Crossing the River**.

Curious	★★★
Connect	★★★
Collaborate	★★★
Communicate	★★
Create	★★
Consider	★★

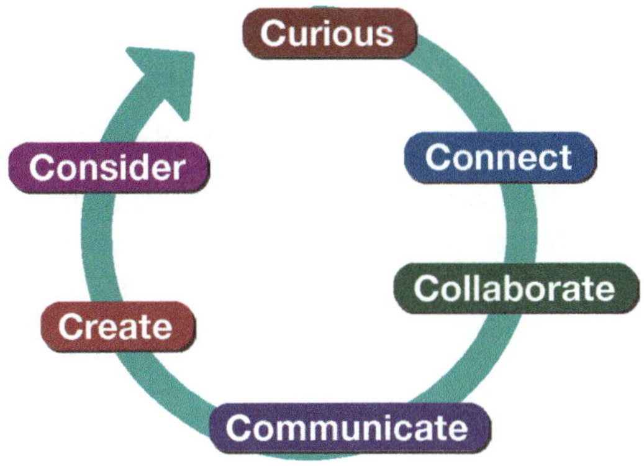

Problem-solving heuristics

The grid shows which problem-solving strategies children are likely to use when investigating the **Connect** tasks in **Crossing the River**. These strategies are further exemplified in the **Connect** stage below (pages 81–84).

Draw a diagram	✓
Make a table / spot a pattern	✓
Make a model / act it out	✓
Try then improve	
Create a systematic list	
Solve a simpler but related problem	
Work backwards	

Key mathematical ideas

Children will discover patterns in the number of jumps it takes Freddie to cross the river. The rule is that Freddie always starts on the first lily pad (not on the left-hand river bank).

If Freddie is able to jump over 1 lily pad (this is the suggested starting point for Year 1 children), then the number of lily pads divided by 2 tells us how many jumps he needs to make. If the number of lily pads is 1 less than a multiple of 2, the number of jumps will need to be rounded up.

For example:

- to cross a river with 6 lily pads, 3 jumps are needed ($6 \div 2 = 3$);
- to cross a river with 5 lily pads, 3 jumps are also needed ($5 \div 2 = 2.5$ which would need to be rounded up to 3 jumps).

It would not be appropriate for children to record this as shown here, and so this should be explored through acting out the scenario and making models.

If Freddie is able to jump over 2 lily pads (this could be the starting point for Year 2 children), then the number of lily pads divided by 3 tells us how many jumps he needs to make. If the number of lily pads is 1 or 2 less than a multiple of 3, this will need to be rounded up.

For example:

- to cross a river with 6 lily pads, 2 jumps are needed ($6 \div 3 = 2$);
- to cross a river with 5 lily pads, 2 jumps are also needed ($5 \div 3 = 1.66$ which would need to be rounded up to 2 jumps);
- to cross a river with 7 lily pads, 3 jumps are needed ($7 \div 3 = 2.33$ which would need to be rounded up to 3 jumps).

Answers to the tasks are given on page 86.

Teaching guidance

If you are using the **Express Pathway**, please turn to pages 81–85.

Teaching guidance for the **Full 6C Pathway** is given below.

The **Teacher's Presentation** PowerPoint file has slides specifically mentioned in this teaching guidance. Some of the slides include notes (within the PowerPoint file) providing 'stumble support' suggestions, that may help you to support children in their problem-solving thinking and enable access for children who are less confident.

Curious

Awe and wonder

Slide 1

Display slide 1.

Ask children to share what they are noticing or wondering.

- I can see a frog and some lily pads.
- I can see 6 lily pads.
- I have noticed that the frog is sitting on the first lily pad on the left.

Develop and deepen

Encourage children to use the words in the word bank to develop and deepen their thoughts and questions.

jump	**starting place**	**how many**	**more lily pads**
over	**across**	**over 1 lily pad**	**fewer lily pads**

Possible lines of enquiry:

- I wonder if the frog has to cross the river to the other side.

Possible lines of enquiry linked to the word bank:

- I wonder if the frog can jump over lily pads, or if he can only jump from one to the next.
- I wonder how many jumps the frog needs to do to cross the river.
- I wonder how many jumps he would need to do if there were more / fewer lily pads.
- I wonder what is the smallest number of jumps the frog needs to do to reach the other side.

Possible lines of enquiry that children may raise now, but should be held back until the **Create** task:

- I wonder what would happen if the frog could jump over two lily pads instead of one.
- I wonder what would happen if the frog started from the bank and not the first lily pad.
- I wonder how many lily pads there could be if Freddie makes *x* jumps.

Consider whether / how to collate and display all the suggested questions, and whether / how to steer children towards selecting an appropriately challenging question.

At this point, explain that Freddie the frog can jump over one lily pad and starts from the first lily pad.

Year 2 children could investigate jumping over one lily pad before considering what happens when Freddie can jump over two lily pads.

- *How can we work out how many jumps Freddie must do?*
- *Now that you know the 'rules', is there anything else you are wondering?*

(The last three lines of enquiry listed above are unlikely to be suggested until the rules are explained.)

Pupil Resources for Full and Express Pathways

Distribute the **Pupil Resources** for **Crossing the River** to the children to help support them in their exploration of the different tasks. Use either of the **Curious Pupil Resources** depending on whether you are following the **Full Pathway** or the **Express Pathway**. The **Pupil Resources** for the remaining 5Cs are common to both pathways.

To provide additional flexibility, each of the **Pupil Resources** is also provided in the **Teacher's Presentation** PowerPoint file (see slides 2–5), so you can display these at any time if you wish.

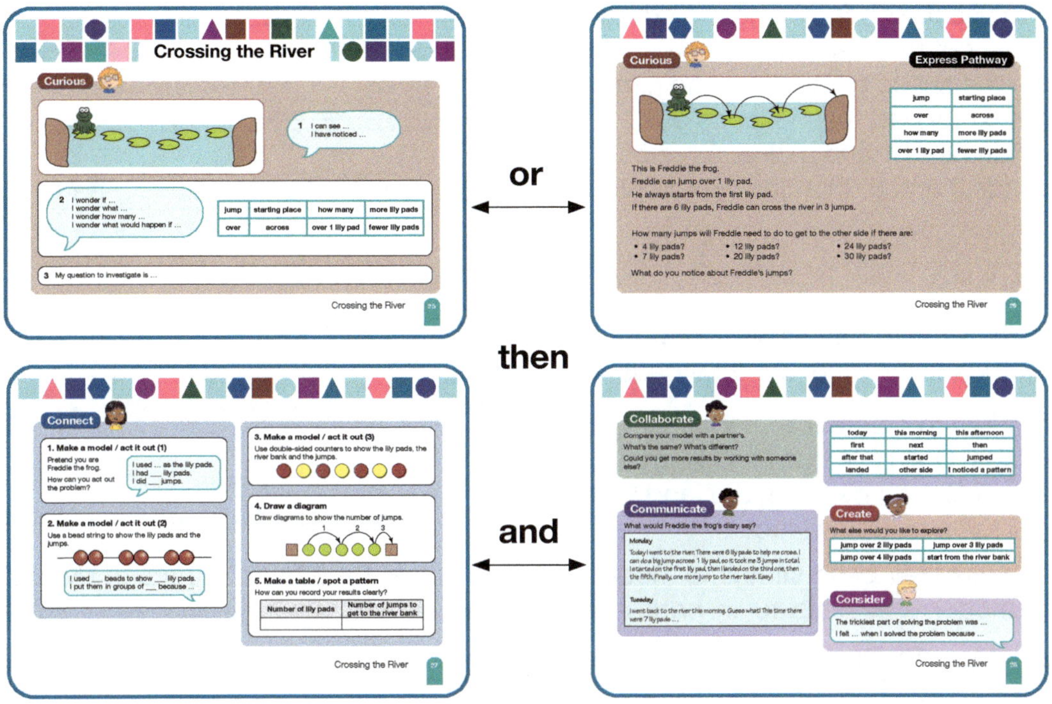

Express Pathway

If you are following the **Express Pathway**, spend a few minutes discussing the **Curious Pupil Resource** with the children. (This is slide 3, so you could display it if you wish.) Use some of the prompts on the previous page to help.

Five tasks involving different strategies are suggested for children to make connections.

1. Make a model / act it out (1)

Additional resources required: spot markers, small gym mats or chalk

Slide 6

Display slide 6.

Encourage children to act out the problem, representing the lily pads using spot markers, small gym mats or chalk markings.

Encourage children to describe the jumps they make. This will support them to see the pattern for even and odd numbers of lily pads.

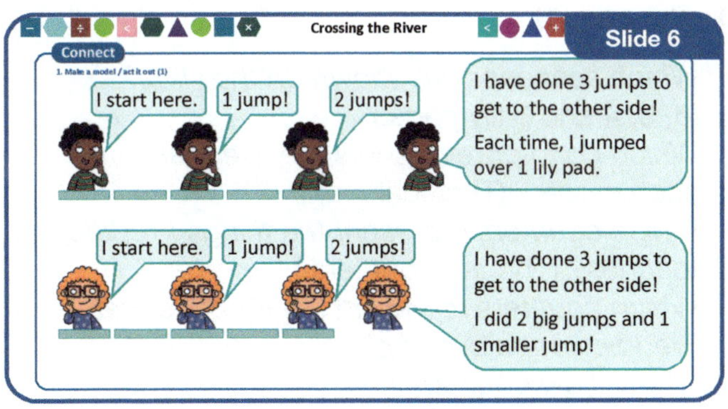

- *Have you noticed anything about the types of jumps that you have described?*
- When the number of lily pads was odd, we had to do big jumps and then one small jump.

2. Make a model / act it out (2)

Additional resources required: bead strings

Slide 7

Display slide 7.

- *How does this bead string show Freddie's jumps?*

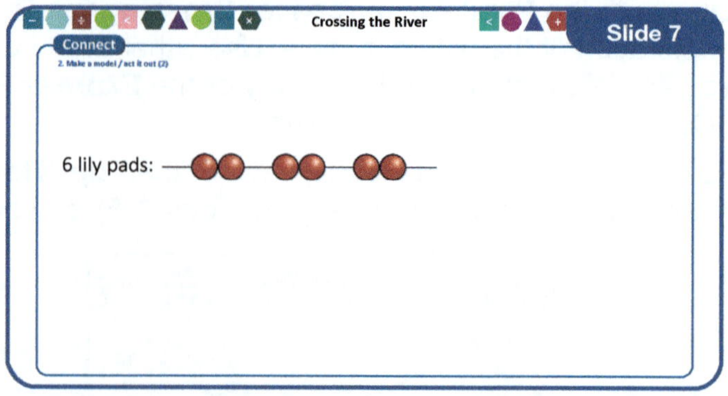

Ask children to use their bead strings to copy the model. Draw out that each bead represents 1 lily pad, with the spacing indicating the jump and then the next lily pad that he lands on. Freddie starts on bead 1 and makes 3 jumps, landing on beads 3 and 5, then the river bank.

Slide 8

Display slide 8.

- *How does the second bead string show Freddie's jumps when there are 9 lily pads?*

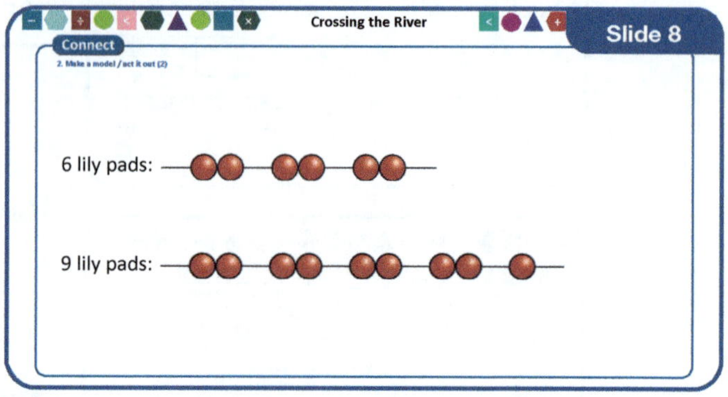

Again, ask children to create this model using their bead strings. Draw out that Freddie makes 4 'big' jumps, landing on beads 3, 5, 7 and 9, then one small jump (5 jumps in total).

Ask children to use their bead strings to model Freddie's jumps for other numbers of lily pads. For example, this model represents 19 lily pads and 10 jumps.

For Year 2 children exploring what happens when Freddie jumps over 2 lily pads at a time, this model represents 23 lily pads and 8 jumps (Freddie starts on bead 1 then lands on beads 4, 7, 10, 13, 16, 19 and 22, with a final jump to the river bank.)

3. Make a model / act it out (3)

Additional resources required: double-sided counters (or counters in two colours)

Slide 9

Display slide 9.

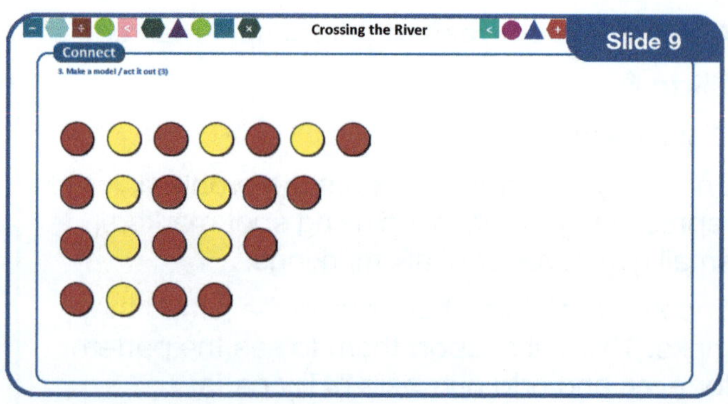

- *Which scenarios are shown on the slide?* (6, 5, 4 and 3 lily pads.)
- *Which counter represents the river bank in each model?* (The final one.)
- *Which counters represent the things Freddie touches?* (The red counters.)
- *Which counters represent the lily pads that Freddie jumps over without touching?* (The yellow counters.)

6 lily pads: 

Ask children to use their double-sided counters to model Freddie's jumps for other numbers of lily pads. For example, this model represents 7 lily pads and 4 jumps.

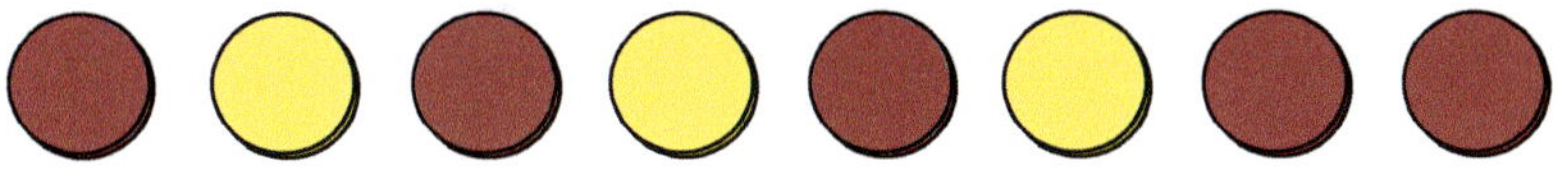

For Year 2 children exploring what happens when Freddie jumps over 2 lily pads at a time, this model represents 8 lily pads and 3 jumps.

4. Draw a diagram

Slide 10

Display slide 10.

Draw out that there are 10 lily pads and the first jump must start from the first lily pad. Model how to draw the remaining three jumps. Agree that there are 5 jumps in total.

Ask children to draw their own simple diagrams to show the jumps for different numbers of lily pads.

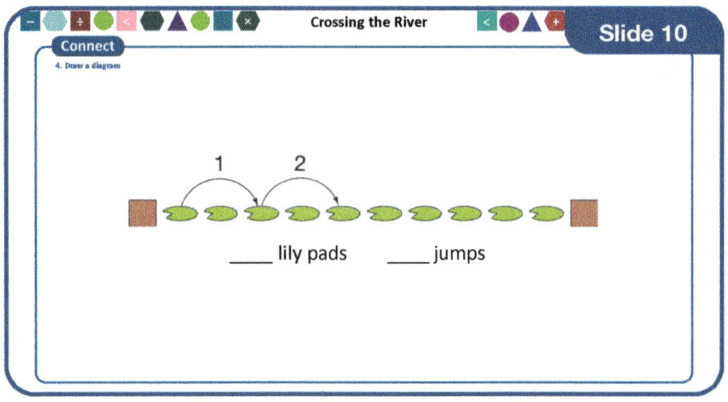

Slides 11 and 12

Alternatively, you could provide children with pre-prepared diagrams for them to annotate. Print slides 11 and 12 and distribute to children as needed.

Children could place their concrete models on sheets of paper and annotate them.

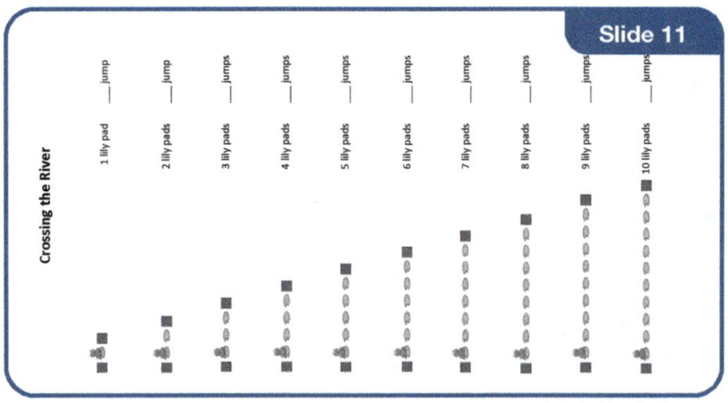

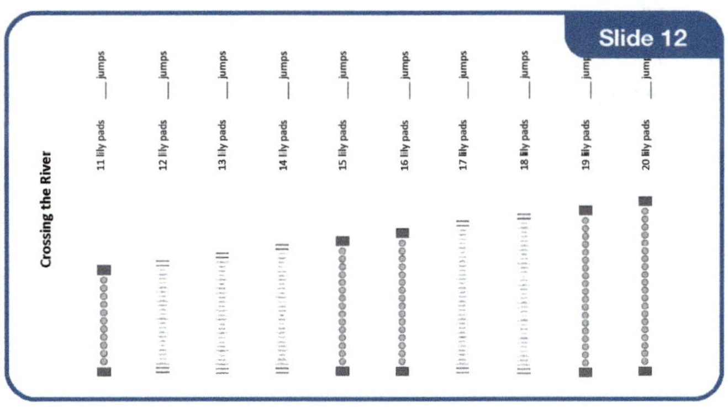

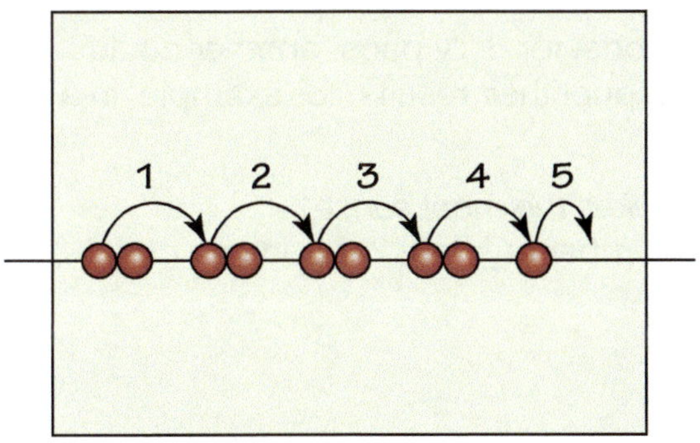

- *I can see that you have drawn a diagram showing that 9 lily pads would need 5 jumps. Could your diagram help you to predict the jumps for a different number of lily pads?*

Connect

5. Make a table / spot a pattern

Once children have explored other strategies, they should be encouraged to present their results in a table.

Slide 13

Display slide 13.

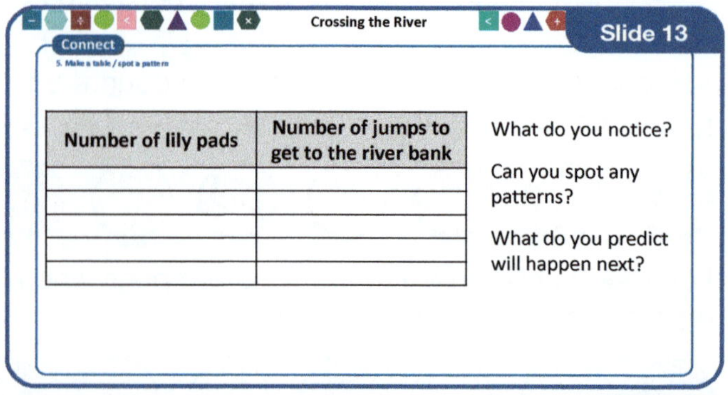

Slide 14

If you would like to provide a table template, print slide 14 and distribute to children.

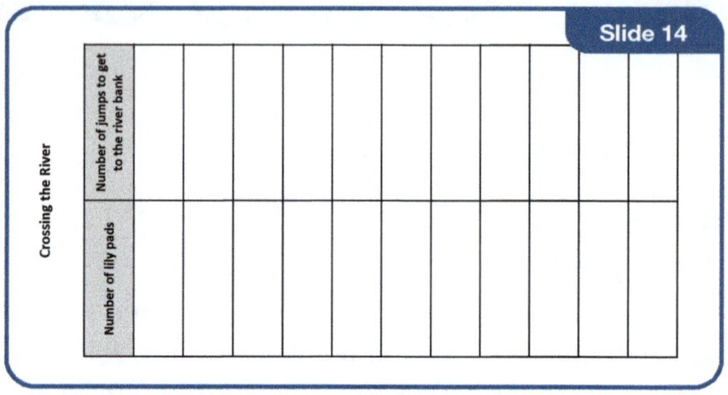

- *What order will you work in?* (I could start with 1 lily pad, then 2, then 3.)
- *Can you spot any patterns?*
- I have noticed that the number of jumps goes up in 1s each time, but it stays on the same number twice: 1, 1, 2, 2, 3, 3, 4, 4. I think 11 lily pads will be 6 jumps and so will 12 lily pads.
- The number of jumps goes up after an even number. So 6 is even and is 3 jumps, but 7 is odd and is 4 jumps. 8 is even and also 4 jumps. 9 is odd and it goes to 5 jumps.
- If you divide the even numbers by 2 you get the number of jumps! 10 divided by 2 is 5 jumps so 12 divided by 2 will be 6 jumps.

Collaborate

Refer to the **Collaborate** task of the **Pupil Resource**.

This task provides an opportunity for children to work in pairs or in small groups to act out the scenarios and to create their models. Each child could build a specific model individually, then compare it with the models made by others in their group.

- *What's the same? What's different?*

Alternatively (or in addition), one child could act out / build a model for 6 lily pads, another could do the same for 7 lily pads, and so on. The children could then pool their results, for example, in a shared table.

- *Whose results will you write in the table first? What order makes the most sense?*
- *How can you be sure that everyone has counted the right number of jumps? Could you check each other's work?*

Children could also work together to write their diary entries.

Refer to the Communicate task of the **Pupil Resource**.

Slide 15

Display slide 15.

This task provides an opportunity for children to write a series of short diary entries in the role of Freddie the frog. They should include information about how many lily pads he encounters each day and how many jumps he must do to cross the river. If children have come to a generalisation and can work out any given number of lily pads, this could also be included.

Alternatively, children could act out Freddie's diary as if they were writing in it.

today	**this morning**	**this afternoon**	**first**	**next**	**then**
after that	**started**	**jumped**	**landed**	**other side**	**I noticed a pattern**

Refer to the Create task of the **Pupil Resource**.

- *Were there any questions asked at the start of the task that you now want to think about?*
- *Is there anything new that you are wondering now?*

Children can use the word bank to prompt further questions of their own.

jump over 2 lily pads	**jump over 3 lily pads**
jump over 5 lily pads	**start from the river bank**

- I wonder what would happen if the frog could jump over two lily pads instead of one.
- I wonder what would happen if the frog started from the bank and not the first lily pad.
- I wonder how many lily pads there could be if Freddie makes x jumps.

Refer to the Consider task of the **Pupil Resource**.

Encourage children to think about the strategies that they have used. The prompts below may help get children started.

Cognitive learning domain

- The trickiest part of solving the problem was …
- I learned …

Affective domain

- I thought that … was fun. I enjoyed it because …
- I found / did not find this problem interesting because …
- I felt … when I solved the problem because …

Metacognitive domain

- I got stuck when … I unstuck myself by …
- I used … to help me. It was helpful because …

Answers

When Freddie's starting point is the first lily pad:

Number of lily pads	Number of jumps to get to the river bank: Jumping over 1 lily pad (steps of 2)	Jumping over 2 lily pads (steps of 3)	Jumping over 3 lily pads (steps of 4)	Jumping over 4 lily pads (steps of 5)
1	1	1	1	1
2	1	1	1	1
3	2	1	1	1
4	2	2	1	1
5	3	2	2	1
6	3	2	2	2
7	4	3	2	2
8	4	3	2	2
9	5	3	3	2
10	5	4	3	2
11	6	4	3	3
12	6	4	3	3
...				
20	10	7	5	4
...				
24	12	8	6	5
...				
30	15	10	8	6

Numbers in blue indicate that only 'big' jumps are needed

When Freddie's starting point is the river bank:

Number of lily pads	Number of jumps to get to the river bank: Jumping over 1 lily pad (steps of 2)	Jumping over 2 lily pads (steps of 3)	Jumping over 3 lily pads (steps of 4)	Jumping over 4 lily pads (steps of 5)
1	1	1	1	1
2	2	1	1	1
3	2	2	1	1
4	3	2	2	1
5	3	2	2	2
6	4	3	2	2
7	4	3	2	2
8	5	3	3	2
9	5	4	3	2
10	6	4	3	3
11	6	4	3	3
12	7	5	4	3

Numbers in blue indicate that only 'big' jumps are needed

Golden Crowns

Year 2

Introduction

Golden Crowns presents a problem that has more than one possible solution. Children will apply their knowledge of multiples of 2 and 5 to help Pam the Pirate buy a new hat. Pam only has two types of coin: 2-crown coins and 5-crown coins. Fortunately, she has lots of each type of coin. Unfortunately, pirate shopkeepers do not give change, so Pam has to make sure that she only pays with coin combinations that total exactly 27 gold crowns.

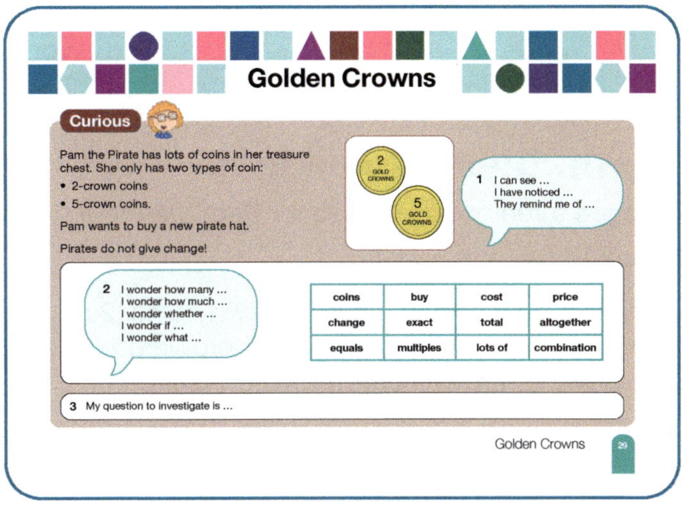

The task will allow children to experience how systematic recording helps them stay in control of their explorations, spot patterns and ensure that all parts of the pattern are present. If children start by using a more random approach, they will need support to be more organised.

This task promotes greater-depth learning in a range of ways, supporting children to identify and articulate underlying mathematical structures. For example, in the Consider stage, greater depth is promoted by supporting children to reflect on learning behaviours that were and were not useful.

This **Teacher Guide** is supported by photocopiable **Pupil Resources** and **Teacher's Presentation** PowerPoint slides for whole-class or group display.

National Curriculum

Golden Crowns covers these Year 2 National Curriculum objectives.

Number and place value

- Count in steps of 2, 3, and 5 from 0, and in tens from any number, forward and backward.

Multiplication and division

- Recall and use multiplication and division facts for the 2, 5 and 10 multiplication tables, including recognising odd and even numbers.

The 6C cycle

The grid on the right shows the relative weighting given to each of the 6 Cs in **Golden Crowns**.

Curious	★★
Connect	★★
Collaborate	★★★
Communicate	★★
Create	★★
Consider	★★★

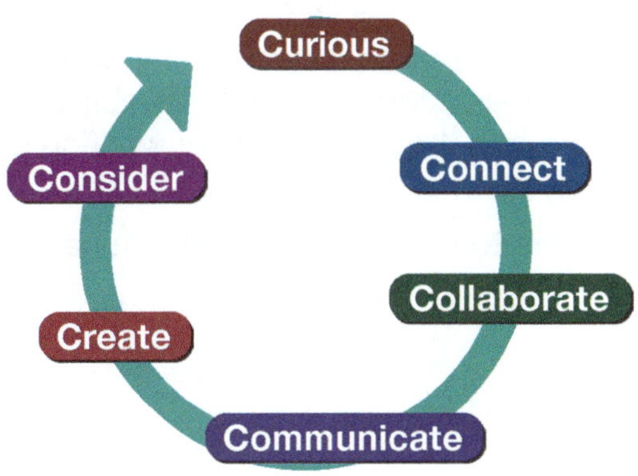

Problem-solving heuristics

The grid shows which problem-solving strategies children are likely to use when investigating the **Connect** tasks in **Golden Crowns**. These strategies are further exemplified in the **Connect** stage below (pages 91–93).

Draw a diagram	
Make a table / spot a pattern	✓
Make a model / act it out	✓
Try then improve	✓
Create a systematic list	✓
Solve a simpler but related problem	
Work backwards	

Key mathematical ideas

There are three possible ways for Pam the Pirate to pay for her new hat without needing change.

She could pay exactly with:

A. one 2-crown coin and five 5-crown coins (2 crowns + 25 crowns);
B. or six 2-crown coins and three 5-crown coins (12 crowns + 15 crowns);
C. or eleven 2-crown coins and one 5-crown coin (22 crowns + 5 crowns).

2-crown coins		5-crown coins	
Number of coins	**Value of coins**	**Number of coins**	**Value of coins**
1	2 gold crowns **(A)**	1	5 gold crowns **(C)**
2	4 gold crowns	2	10 gold crowns
3	6 gold crowns	3	15 gold crowns **(B)**
4	8 gold crowns	4	20 gold crowns
5	10 gold crowns	5	25 gold crowns **(A)**
6	12 gold crowns **(B)**		
7	14 gold crowns		
8	16 gold crowns		
9	18 gold crowns		
10	20 gold crowns		
11	22 gold crowns **(C)**		
12	24 gold crowns		
13	26 gold crowns		

By listing the values of multiple coins in a table, children will be able to more easily keep track of the different totals possible.

Through using the recording strips that are introduced in the first **Connect** task, children can easily move calculations around to establish a logical order. This will help them to identify patterns and understand the importance of working systematically. The patterns are important, as it is through closely examining these that children will see whether they have all possibilities or not.

Answers to the tasks are given on page 95.

Teaching guidance

If you are using the **Express Pathway**, please turn to pages 90–94.

Teaching guidance for the **Full 6C Pathway** is given below.

The **Teacher's Presentation** PowerPoint file has slides specifically mentioned in this teaching guidance. Some of the slides include notes (within the PowerPoint file) providing 'stumble support' suggestions, that may help you to support children in their problem-solving thinking and enable access for children who are less confident.

Awe and wonder

Slide 1

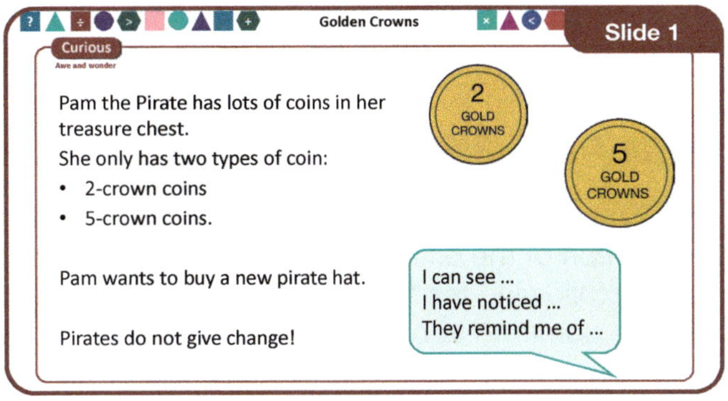

Display slide 1.

Ask children to share what they are noticing or wondering.

- I can see that the 2-crown coin is smaller than the 5-crown coin.
- I have noticed that Pam has lots of coins.
- I can see that she wants to buy a new hat, but we don't know how much it costs.
- I can see that pirates don't give change, so Pam will need to pay with exactly the right amount.
- They remind me of the coins that we use: we have a 2 pence coin and a 5 pence coin.

Develop and deepen

- Pam wants to buy a new hat, but she needs to pay only the exact price, because she won't be given any change.

Encourage children to use the words in the word bank to develop and deepen their thoughts and questions.

coins	**buy**	**cost**	**price**
change	**exact**	**total**	**altogether**
equals	**multiples**	**lots of**	**combination**

Check whether all children understand the meaning of 'change' and 'combinations', explaining if needed.

Possible lines of enquiry before children know how much the hat costs:

- I wonder how many coins Pam has in total.
- I wonder whether she will have enough coins to buy the hat.
- I wonder how much the hat costs.

Explore with the children different prices that the hat could and could not be, if it must be possible for Pam to pay and receive no change. For example:

- *If the hat cost 17 gold crowns, how could Pam pay?*
- Pam could pay for the hat with one 2-crown coin and three 5-crown coins.
- *Why couldn't she pay with four 5-crown coins?*
- Four 5-crown coins are worth 20 gold crowns, so Pam would lose 3 gold crowns. So she wouldn't want to do that.

Now tell the children that the hat Pam wants costs **27 gold crowns**.

Possible lines of enquiry once children know how much the hat costs:

- I wonder what combination of coins will total 27 gold crowns.
- I wonder if there is more than one way to make a total of 27 gold crowns.
- I wonder if Pam would be able to buy the hat if it cost 26 gold crown or 28 gold crowns instead.

Possible lines of enquiry that children may raise now, but should be held back until the **Create** task:

- I wonder what other possible totals we could make with 2-crown and 5-crown coins.
- I wonder what totals are impossible with 2-crown and 5-crown coins.
- I wonder if I could make the same totals using 3-crown and 4-crown coins.

Consider whether / how to collate and display all the suggested questions, and whether / how to steer children towards selecting an appropriately challenging question.

Pupil Resources for Full and Express Pathways

Distribute the **Pupil Resources** for **Golden Crowns** to the children to help support them in their exploration of the different tasks. Use either of the **Curious Pupil Resources** depending on whether you are following the **Full Pathway** or the **Express Pathway**. The **Pupil Resources** for the remaining 5Cs are common to both pathways.

To provide additional flexibility in the classroom, each of the **Pupil Resources** is also provided in the **Teacher's Presentation** PowerPoint file (see slides 2–5), so you can display these at any time if you wish.

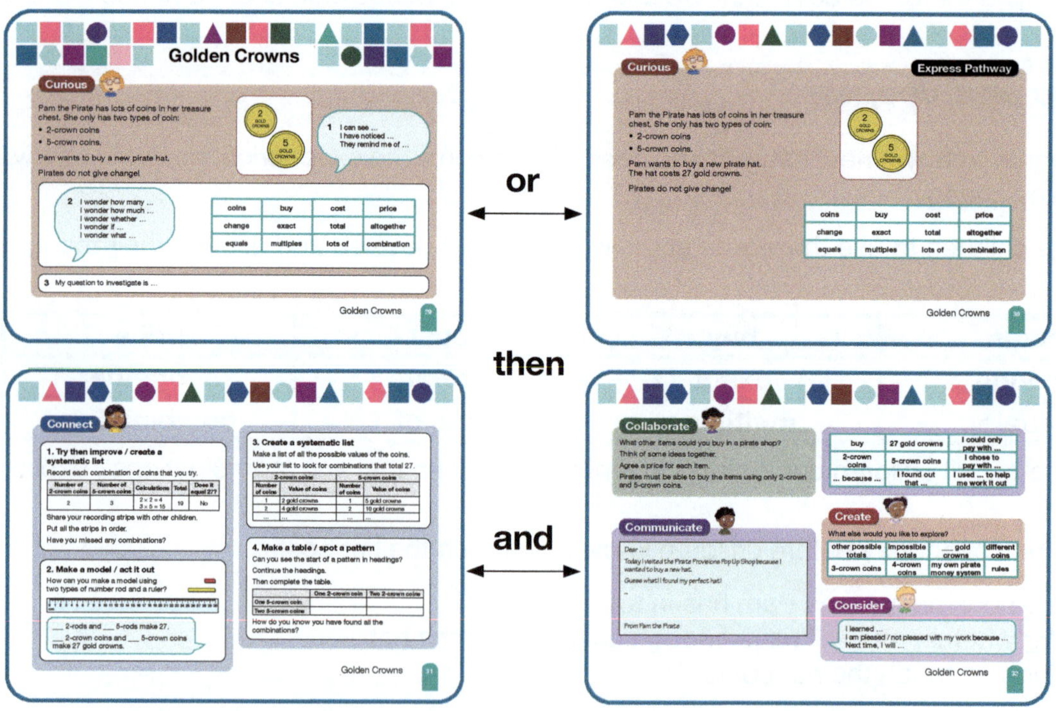

Express Pathway

If you are following the **Express Pathway**, spend a few minutes discussing the **Curious Pupil Resource** with the children. (This is slide 3, so you could display it if you wish.) Use some of the prompts on the previous page to help.

Four tasks involving different strategies are suggested for children to make connections.

1. Try then improve / create a systematic list

Slide 6

Print slide 6 and distribute to children. Ask them to cut out the individual strips and use them to record their attempts.

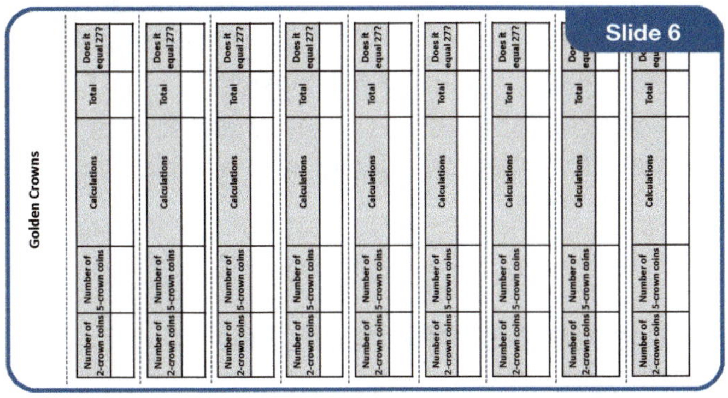

Slide 7

Display slide 7, to give children an example of how to complete the recording strips.

The completed strips could then be collated into a larger set. Children could work together to look for patterns and rearrange the strips into a systematic order. This should allow them to see if they have missed any combinations.

- *How did you decide on this order?*
- *Now that the strips are in a systematic order, what do you notice?*

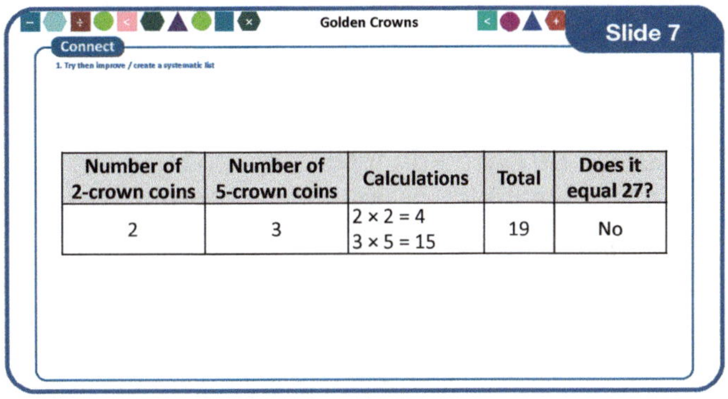

2. Make a model / act it out

Additional resources required: number rods, 30 cm rulers

Slide 8

Display slide 8.

This is an ideal entry method for the children to understand the problem. However, they may need support to become systematic so that they are confident they have all the possible solutions.

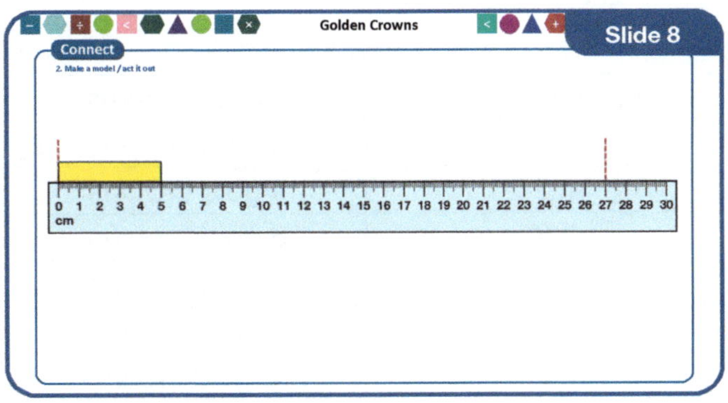

- *Which number rods could you use to represent the problem?*
- We could use 2-rods to represent the 2-crown coins, and 5-rods to represent the 5-crown coins.
- *Number rods are an exact number of centimetres long. How could you use a ruler to help you with your models?*
- We need a total of 27, so if we put the rods in a row, the row needs to be 27 centimetres.
- *How many different rows of 2-rods and 5-rods can you make that are 27 centimetres long?*

Slide 9

Print slide 9 and distribute to the children to use as a recording sheet. They can divide the bottom bar of each pair to show the number of 2-rods and 5-rods used, then complete the sentences with the numbers of rods.

Provide additional copies of the sheet so that children do not automatically stop after finding three solutions.

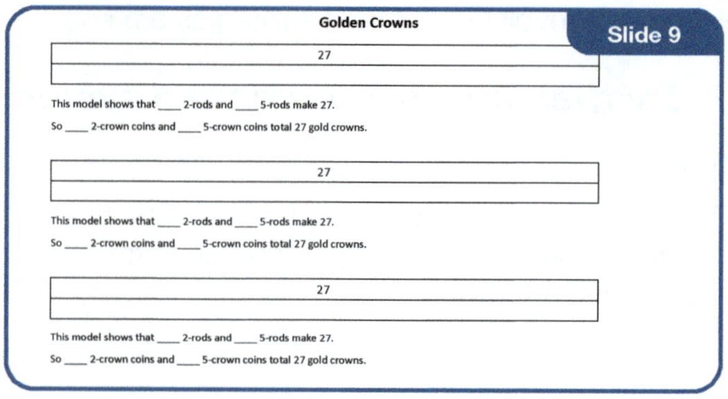

3. Create a systematic list

Slide 10

Display slide 10.

Some children may choose to create a systematic list of all the different coin values. This could be based on their knowledge of the 2 and 5 times tables or counting patterns. They could make a table similar to the one on the slide.

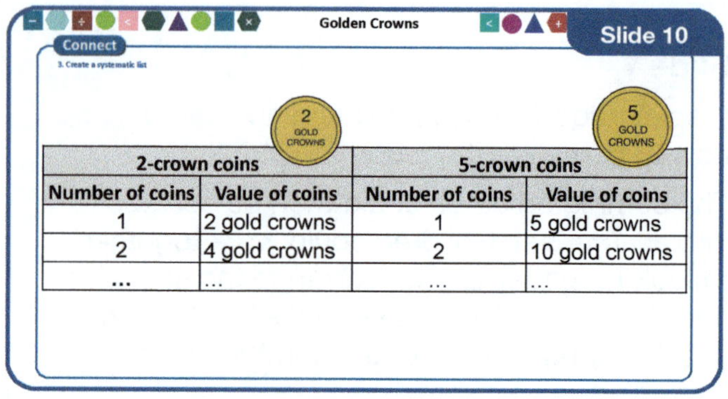

- *Why do we only need to go up to thirteen 2-crown coins?* (Because fourteen 2-crown coins would be worth 28 gold crowns, which is more than the price of the hat.)
- *Why do we only need to go up to five 5-crown coins?* (Because six 5-crown coins would be worth 30 gold crowns, which is too much.)

Children can then use their completed list to search systematically for combinations that total 27.

For example, they could start with the first value in the '2-crown coins' part of the table (2 gold crowns), and go through each of the values in the '5-crown coins' part of the table in turn, looking for a combination that works.

2-crown coins		5-crown coins	
Number of coins	**Value of coins**	**Number of coins**	**Value of coins**
1	2 gold crowns	1	5 gold crowns
2	4 gold crowns	2	10 gold crowns
3	6 gold crowns	3	15 gold crowns
4	8 gold crowns	4	20 gold crowns
5	10 gold crowns	5	25 gold crowns
6	12 gold crowns		
7	14 gold crowns		
8	16 gold crowns		
9	18 gold crowns		
10	20 gold crowns		
11	22 gold crowns		
12	24 gold crowns		
13	26 gold crowns		

4. Make a table / spot a pattern

This could be used as a next step after creating a systematic list as described above, or as a separate strategy.

Slide 11

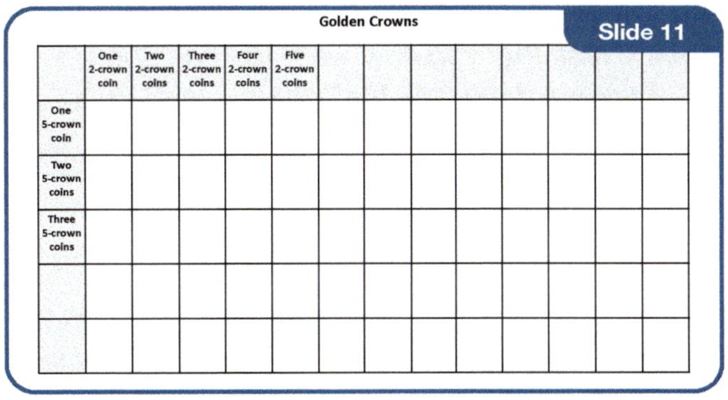

Print slide 11 and distribute to the children.

- *Can you see the pattern in the row and column headings?*
- *Can you complete the headings?*

Then encourage children to populate the cells.

- *What would you write in the cell for 'One 2-crown coin' and 'One 5-crown coin'?*
- I would write: '2 + 5 = 7'.

When children have completed their tables, ask them to highlight any combinations that total 27.

- *How could this table help you feel confident that you have found all the possible combinations?*

Refer to the **Collaborate** task of the **Pupil Resource**.

This task provides an opportunity for children to collaborate during the first **Connect** task. Once children have completed some recording strips, they should team up in a group to collate all their strips. They can then discuss logical ways to place all their strips in order.

Ahead of the **Communicate** task, children could work in pairs to come up with ideas for additional items that could be sold in the pirate shop, that can also be bought using only 2-crown and 5-crown coins.

Refer to the **Communicate** task of the **Pupil Resource**.

Slide 12

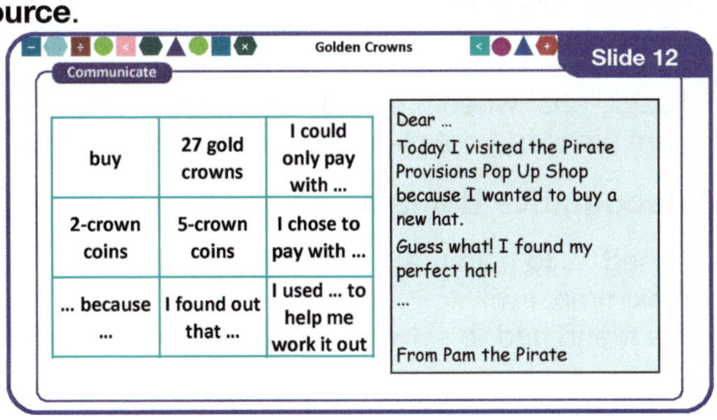

Display slide 12.

This task provides an opportunity for the children to write a letter to a friend about their visit to the 'Pirate Provisions Pop Up Shop'. Within the letter, children should use what they have found out to explain the different ways that they could pay for the pirate hat. They could then say which method of payment they chose and why.

They could also list additional items in the shop that can also be bought using only 2-crown and 5-crown coins, as discussed in pairs in the **Collaborate** task.

Create

Refer to the **Create** task of the **Pupil Resource**.

Slide 13

Display slide 13.

- *Were there any questions asked at the start of the task that you now want to think about?*
- *Is there anything new that you are wondering now?*

Children can use the word bank to prompt further questions of their own.

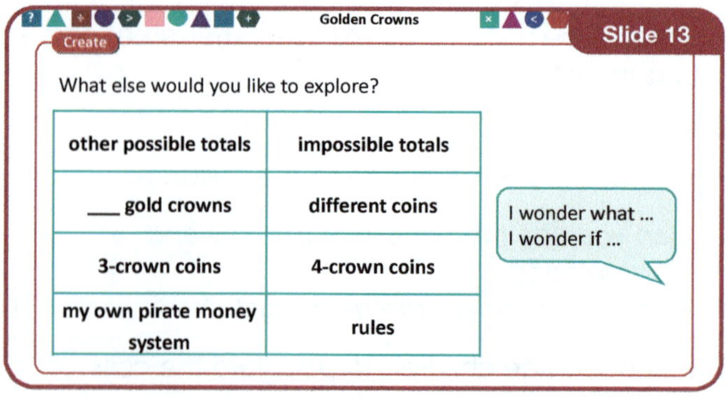

other possible totals	**impossible totals**	**___ gold crowns**	**different coins**
3-crown coins	**4-crown coins**	**my own pirate money system**	**rules**

- I wonder what other possible totals we could make with 2-crown and 5-crown coins.
- I wonder what totals are impossible with 2-crown and 5-crown coins.
- I wonder if I could make the same totals using 3-crown and 4-crown coins.
- I wonder if I could make up my own pirate money. What coins would I have? What totals could I make with them?

Consider

Refer to the **Consider** task of the **Pupil Resource**.

Slide 14

Display slide 14.

Encourage children to think about the strategies that they have used. The prompts below may help get children started.

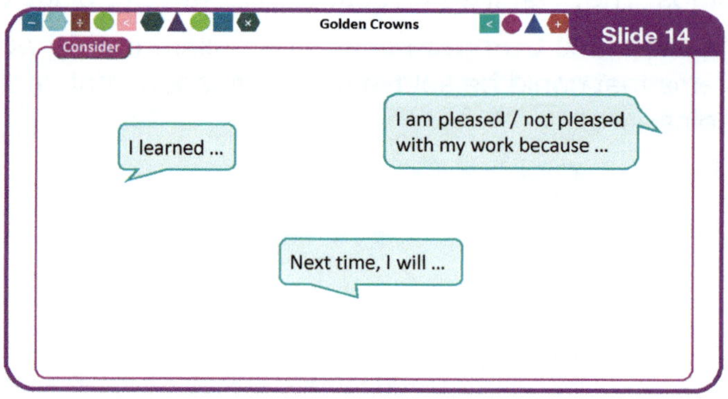

Cognitive learning domain

- The trickiest part of solving the problem was …
- I learned …

Affective domain

- … was fun because …
- I said 'aha' when … I said it because …
- I am pleased / not pleased with my work because …

Metacognitive domain

- I used … to help me. It was helpful because …
- Next time, I will …
- If a friend had to solve this problem, I would give them this tip: …

Answers

There are three possible ways to make 27 by combining 2s and 5s:

A. one 2-crown coin + five 5-crown coins (2 crowns + 25 crowns);
B. or six 2-crown coins + three 5-crown coins (12 crowns + 15 crowns);
C. or eleven 2-crown coins + one 5-crown coin (22 crowns + 5 crowns).

Can I Halve It?

Year 1

Introduction

Can I Halve It? provides children with the opportunity to deepen their ability to recognise and find halves of objects, shapes and quantities. Statements about halving are presented, to promote discussion about children's own experiences of, and thoughts about, sharing equally.

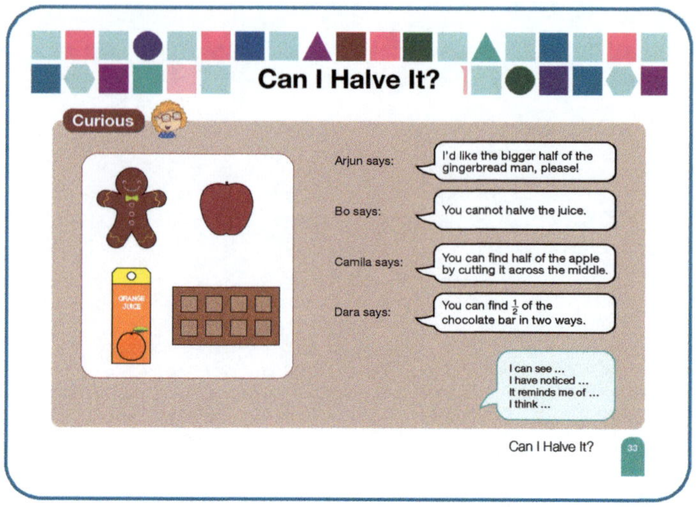

Children will explore the statements, and work out to prove or disprove each one. They will then work together to prepare and present their proof.

Following this, the children should then be encouraged to explore additional items of their choosing in order to provide additional examples. This should lead them to generalise that everything can be halved as long as each of the two groups contain an equal amount.

This task promotes greater-depth learning in a range of ways, supporting children to identify and articulate underlying mathematical structures. For example, in the **Curious** stage, greater depth is promoted by supporting children to offer more than one good mathematical question.

This **Teacher Guide** is supported by photocopiable **Pupil Resources** and **Teacher's Presentation** PowerPoint slides for whole-class or group display.

National Curriculum

Can I Halve It? covers these Year 1 National Curriculum objectives.

Fractions

- Recognise, find and name a half as one of two equal parts of an object, shape or quantity.

The 6C cycle

The grid on the right shows the relative weighting given to each of the 6 Cs in **Can I Halve It?**

Curious	★★★
Connect	★★★
Collaborate	★★
Communicate	★★
Create	★★
Consider	★

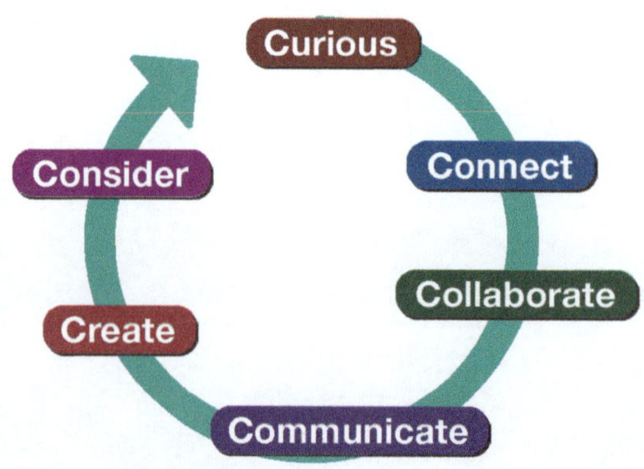

Problem-solving heuristics

The grid shows which problem-solving strategies children are likely to use when investigating the **Connect** tasks in **Can I Halve It?** These strategies are further exemplified in the **Connect** stage below (pages 100–102).

Draw a diagram	✓
Make a table / spot a pattern	
Make a model / act it out	✓
Try then improve	✓
Create a systematic list	
Solve a simpler but related problem	
Work backwards	

Key mathematical ideas

Children may have experienced halves already through everyday language, through exploring turns and when telling the time to half past the hour. By exploring the given statements, children will reinforce their understanding that halves are two equal parts of a whole.

Statement 1: 'I'd like the bigger half of the gingerbread man, please!'

Statement 2: 'You cannot halve the juice.'

These statements have been used to prompt discussions around common misconceptions linked to understanding of fractions in young children. It is a common misconception that halving merely requires splitting something into two parts without the secure understanding that these parts need to be equal. It is also likely that young children may consider that halving requires 'cutting' and if the juice cartoon was cut, then the liquid would spill out.

Statement 3: 'You can find half of the apple by cutting it across the middle.'

This statement prompts discussion around the importance of the direction an object is halved when making two equal pieces. This may also lead children to consider whether there are different ways that they can halve the apple as described in the final statement.

Statement 4: 'You can find $\frac{1}{2}$ of the chocolate bar in two ways.'

This statement has been provided to encourage children to think about different ways that they can halve an object. They may consider halving the 8 pieces of the chocolate bar by splitting the bar into 2 parts with 4 pieces, either horizontally or vertically. They may also consider splitting the bar into 4 parts with 2 pieces and then grouping these into 2 groups. They may also split the bar using a diagonal line and then discuss whether this would be a helpful method. Additionally, they could split the bar into 8 pieces and share them alternating between two groups: A, B, A, B, and so on.

This task will also involve the children having to prove or disprove their own statements as well as those of their classmates. This terminology may not have been introduced yet, so we ask the children to talk about why they think they are correct. To do this they must understand the definition of what a half is and be able to justify that what they have done is within this definition.

Answers are given in situ throughout the tasks because of the cumulative nature of the tasks.

Teaching guidance

Teaching guidance is given below.

The **Teacher's Presentation** PowerPoint file has slides specifically mentioned in this teaching guidance. Some of the slides include notes (within the PowerPoint file) providing 'stumble support' suggestions, that may help you to support children in their problem-solving thinking and enable access for children who are less confident.

Awe and wonder

Reveal slides 1–2 one at a time, allowing time for discussion and observations as prompted below.

Slide 1

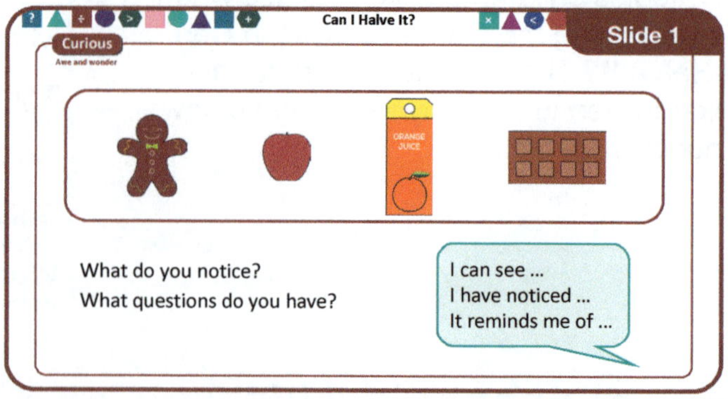

- I can see things that you can eat.
- I have noticed that one of the items is something that you drink, but the others you eat.
- I have noticed that the chocolate bar has 8 smaller pieces.
- I can see that there is one of each item.
- It reminds me of things that might be packed in a picnic.

Slide 2

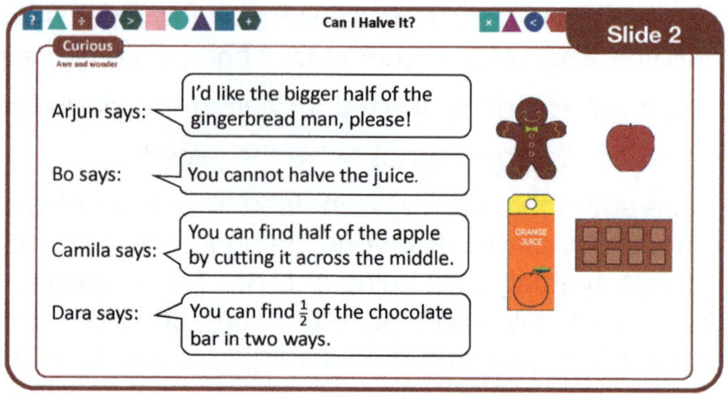

- I can see that each of the children is talking about halves.
- I have noticed that each child matches with a different item.
- I think that some of the children are wrong.
- I can see a fraction.
- I can see the fraction $\frac{1}{2}$.

Develop and deepen

Encourage children to use the words in the word bank to develop and deepen their thoughts and questions.

half	**share**	**equal**	**unequal**
parts	**correct**	**incorrect**	**more than**
less than	**fraction**	$\frac{1}{2}$	**halve**

Possible lines of enquiry linked to the word bank:

- I wonder what a half is.
- I wonder if each of the items can be shared into equal parts.
- I wonder which of the children are correct / incorrect.
- I wonder what I could say to help the children who are incorrect.
- I wonder how I could check what each child has said.
- I wonder if there are more than two ways to halve the chocolate bar.

Possible lines of enquiry that children may raise now, but should be held back until the **Create** task:

- I wonder if you can halve any object.
- I wonder how many different ways you could halve an apple / gingerbread man.
- I wonder which of these could also be split into quarters.

Consider whether / how to collate and display all the suggested questions, and whether / how to steer children towards selecting an appropriately challenging question.

Pupil Resources

Distribute the **Pupil Resources** for **Can I Halve It?** to the children to help support them in their exploration of the different tasks.

To provide additional flexibility in the classroom, each of the **Pupil Resources** is also provided in the **Teacher's Presentation** PowerPoint file (see slides 3–6), so you can display these at any time if you wish.

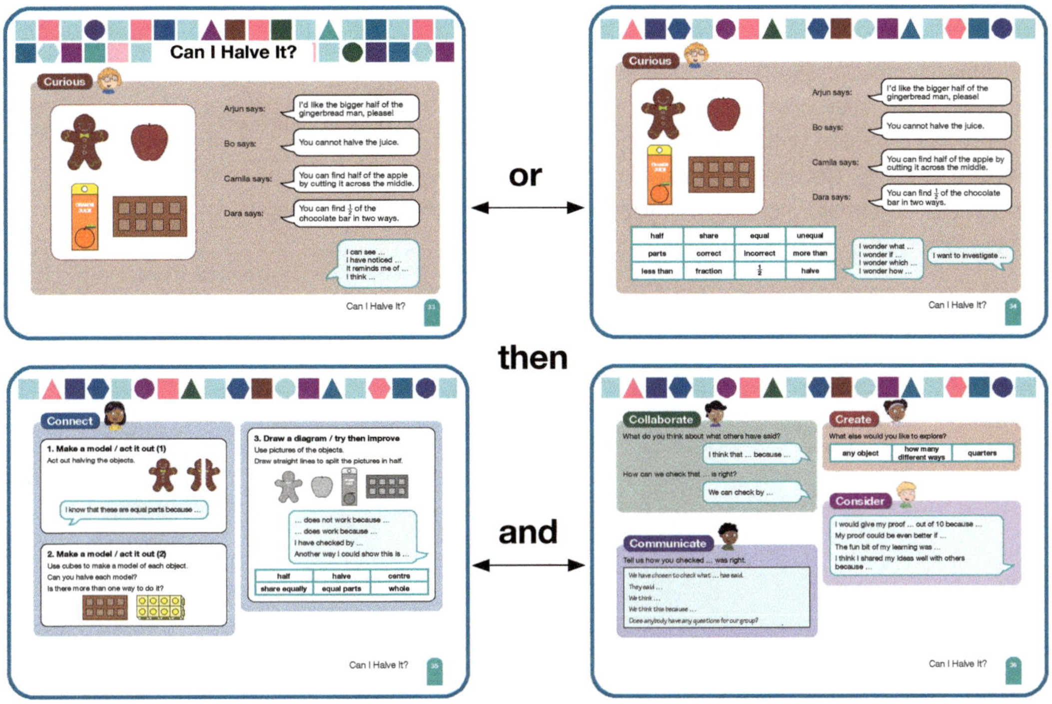

Connect

Three tasks involving different strategies are suggested for children to make connections.

1. Make a model / act it out (1)

Additional resources required: gingerbread men, apples, cartons of orange juice, bars of chocolate with 8 pieces (optional)

Slide 7

Display slide 7.

Encourage children to physically manipulate objects to discover how they can halve the items. Ideally, real-life objects will be used, but otherwise slides 11 and 12 could be printed out for children to cut out and fold / cut up. A jug of water could be used to represent the orange juice.

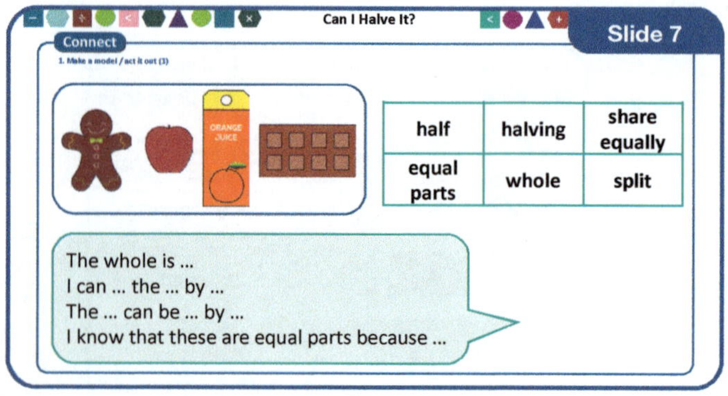

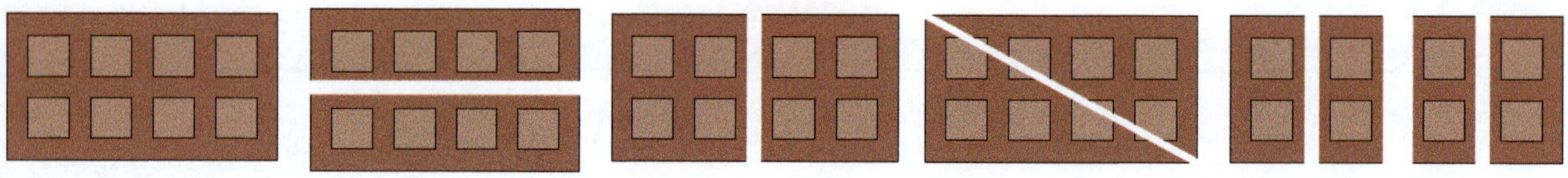

The children should use the sentence stems on slide 7 to help them describe what they are doing and justify how they know they have found a half. They can use the word bank to help fill in the gaps.

If real items are being used, ensure all children are supervised if they are using knives to cut the objects. It may be more appropriate for the children to direct and instruct the adults in the room to cut for them.

2. Make a model / act it out (2)

Additional resources required: interlocking cubes

The children could recreate the items by building their own representations using interlocking cubes. The word bank and speaking frame from slide 7 could be used to support verbal explanations.

Slide 8

Display slide 8, which shows how the chocolate bar could be represented using 8 interlocking cubes.

Ask children to explore different ways to split their 8 cubes into halves.

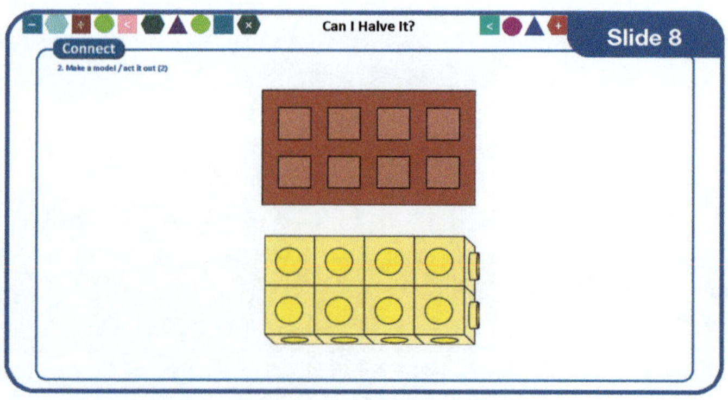

Slide 9

Display slide 9 as an example.

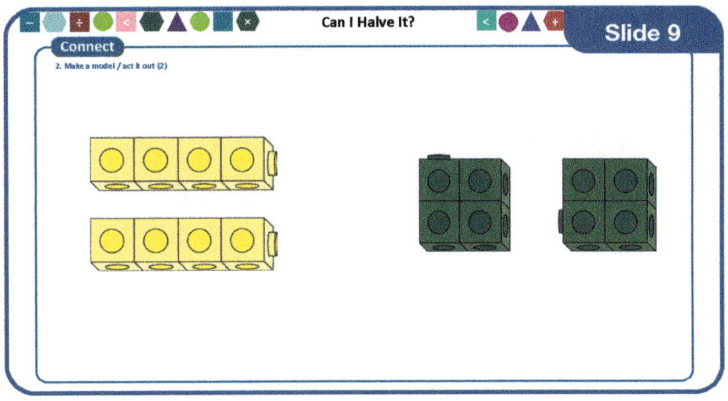

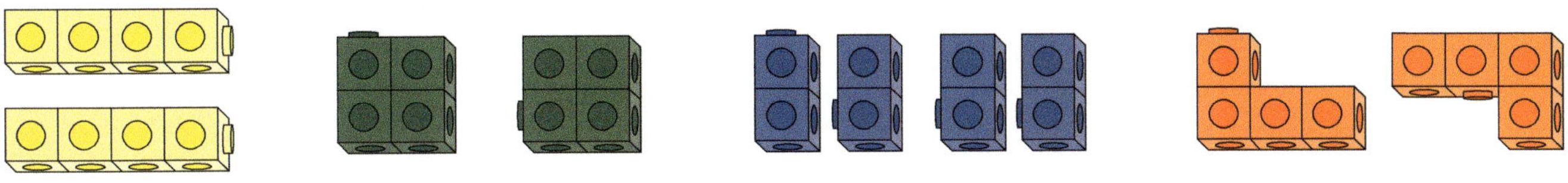

Four different ways to divide eight interlocking cubes, representing the chocolate bar, into halves

- I can halve the chocolate bar by sharing the pieces into 2 groups. The groups are equal because they both have 4 pieces of the chocolate bar. 8 pieces halved is 4 pieces.

Slide 10

Display slide 10.

The gingerbread man will be harder and more interesting for the children to attempt to represent using interlocking cubes. Ask them to first of all copy the model shown on the slide using interlocking cubes.

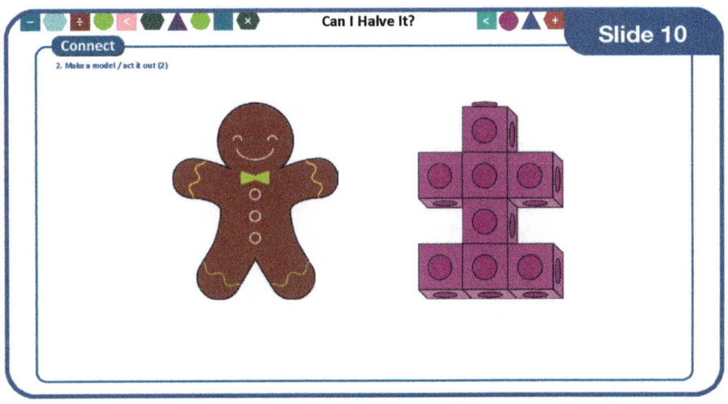

- *Will this model be easy to halve?*
- I could halve my model down the middle, but I would have to break the cubes.
- I could halve it across the middle, and each person would get four cubes. But one person would get the head and arms, and the other would get the legs, so it's not exactly the same.
- *How could you change your model to make it easy to split down the middle?*

Children may add four more cubes, doubling the thickness of the head and body.

- I have made the body thicker so that this time I can halve it down the middle.

- *Is this still a model of the gingerbread man? Does it help solve your problem?* (Yes.)

3. *Draw a diagram / try then improve*

Additional resources required: rulers, scissors

Slides 11 and 12

Print out slides 11 and 12 and distribute to children. Depending on your lesson length, you may decide to select only one or two of the items for this task.

Ask children to use the handouts to draw lines on the objects to show how they could each be shared equally by two people.

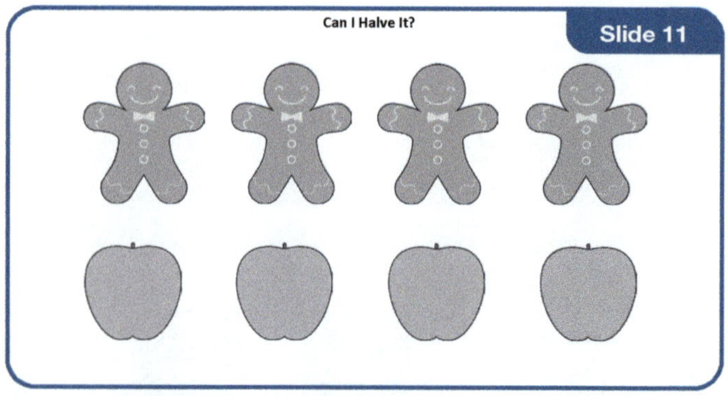

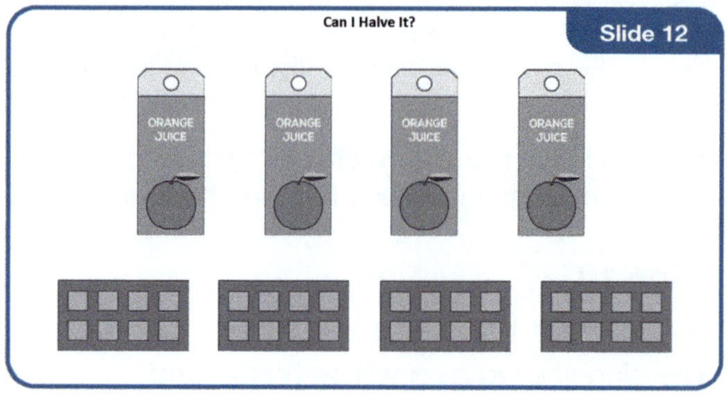

Slide 13

Display slide 13.

Use the sentence stems to encourage children to check their answers and justify their reasoning.

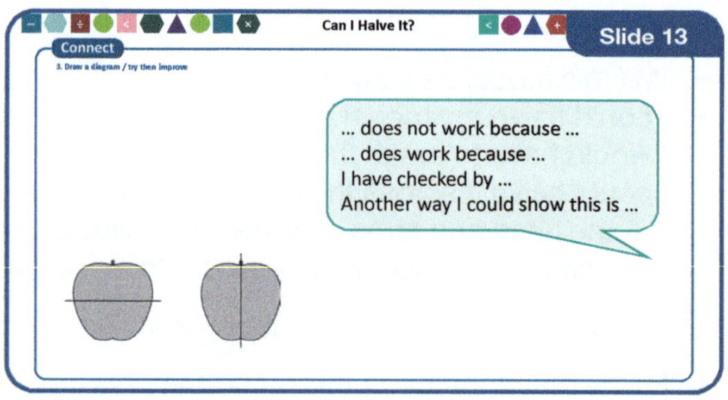

- It does not work if you split the apple across the middle because the top piece is more than the bottom piece. It does work when you split it down the middle though. I have checked by drawing on the pictures. Another way I could show this is if [teacher] cuts the apples in both ways.
- *Would cutting it across the middle work if the apple were a different shape?*
- *Would this work with the juice?*
- *What would be the best way to share it?*
- *What about the gingerbread man?*

Some children might suggest that you could weigh each piece to check whether they are equal. If you are using real objects and have scales available this is a valuable exercise.

Refer to the **Collaborate** task of the **Pupil Resource**.

Slide 14

Display slide 14.

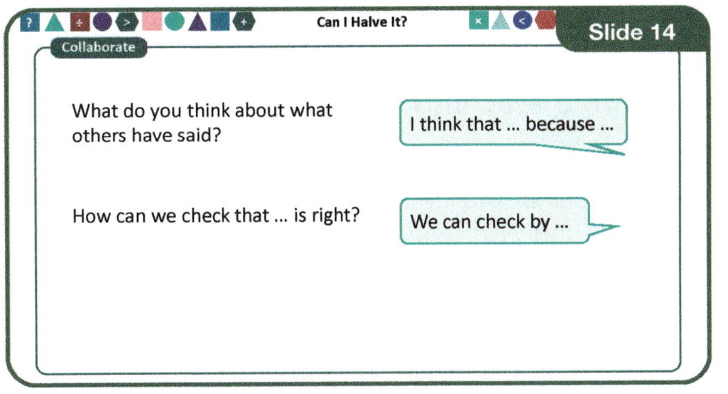

This task provides an opportunity for the children to work in pairs or small groups. First, they can discuss what they think about the statements from each child (Arjun, Bo, Camila and Dara), before testing out their own ideas and forming their proof.

Children should be encouraged to take turns to speak and to ask questions.

- *What do you think about what others have said?*
- I think that … because …
- *How can we check that … is right?*
- We can check by …

Refer to the **Communicate** task of the **Pupil Resource**.

This task provides an opportunity for the children to present their proof as a short presentation.

To provide an engaging purpose for their presentations, you could ask children to imagine that the class is making a television programme called 'We Can Convince You'. Children who are not presenting become audience members. All groups should be encouraged to present their proof for their chosen statement.

Here are some helpful sentence stems:

- We have chosen to check what … has said.
- They said …
- We think …
- We think this because …
- Does anybody have any questions for our group?

Refer to the **Create** task of the **Pupil Resource**.

- *Were there any questions asked at the start of the task that you now want to think about?*
- *Is there anything new that you are wondering now?*

Children can use the word bank to prompt further questions of their own.

any object	**how many different ways**	**quarters**

- I wonder if you can halve any object.
- I wonder how many different ways you could halve an apple / gingerbread man.
- I wonder which of these could also be split into quarters.

Consider

Refer to the Consider task of the **Pupil Resource**.

Encourage children to think about the strategies that they have used. The prompts below may help get children started.

Cognitive learning domain

- I would give my proof … out of 10 because …
- My proof could be even better if …
- By working on this problem, I learned …

Affective domain

- The fun bit of my learning was …
- It was fun because …
- I am pleased / not pleased with my answer because …

Metacognitive domain

- I think I shared my ideas well with others because …
- Next time, I will …

Poppy the Penguin

Year 2

Introduction

Poppy the Penguin gives children an opportunity to extend their experience of finding fractions of amounts in the context of an engaging story. Through exploration of the story, children will consider how the whole diminishes each time Poppy gives some of her fish away, and how this changes depending on the order in which the fractions are used.

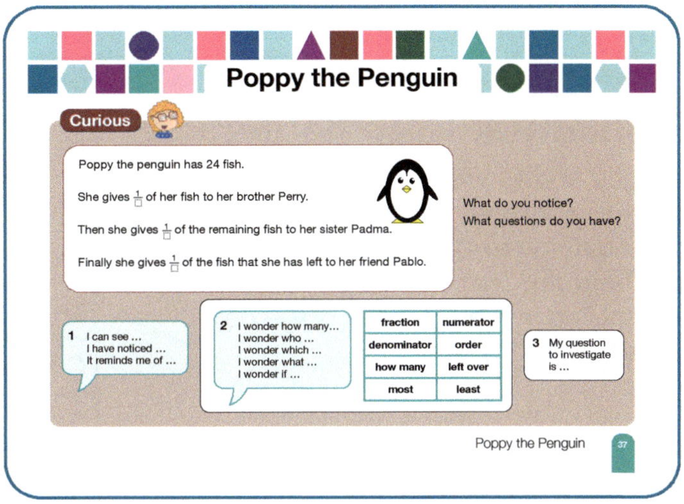

Children make links between finding fractions of amounts, and division (sharing a whole into equal-sized groups). Concrete resources and diagrams such as bar models will be useful representations to show the equality of the parts.

Through exploration of using the fractions in different positions in the story, children will discover what stays the same and what changes. They look for patterns and use models and diagrams to help explain what they notice.

This task promotes greater-depth learning in a range of ways, supporting children to identify and articulate underlying mathematical structures. For example, in the **Connect** stage, greater depth is promoted by supporting children to choose an appropriate strategy, explain their choice and understand that sometimes more than one answer is possible.

This **Teacher Guide** is supported by photocopiable **Pupil Resources** and **Teacher's Presentation** PowerPoint slides for whole-class or group display.

National Curriculum

Poppy the Penguin covers these Year 2 National Curriculum objectives.

Fractions

- Recognise, find and name fractions $\frac{1}{3}$, $\frac{1}{4}$, $\frac{2}{4}$ and $\frac{3}{4}$ of a length, shape, set of objects or quantity.

Multiplication and division

- Solve problems involving multiplication and division, using materials, arrays, repeated addition, mental methods, and multiplication and division facts, including problems in contexts.

The 6C cycle

The grid on the right shows the relative weighting given to each of the 6 Cs in **Poppy the Penguin**.

Curious	★★
Connect	★★★
Collaborate	★★
Communicate	★★
Create	★
Consider	★★

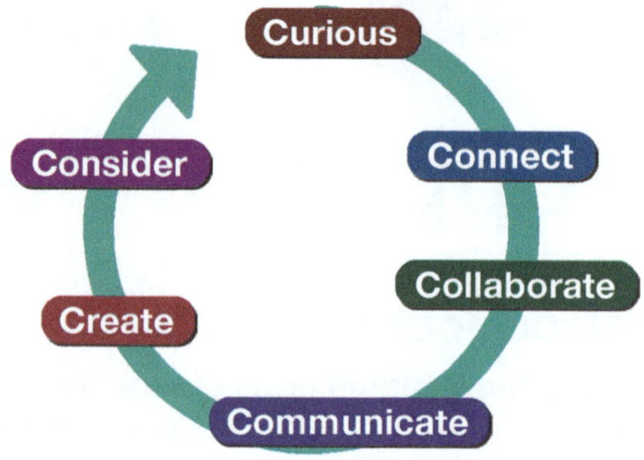

Problem-solving heuristics

The grid shows which problem-solving strategies children are likely to use when investigating the Connect tasks in **Poppy the Penguin**. These strategies are further exemplified in the Connect stage below (pages 109–111).

Draw a diagram	✓
Make a table / spot a pattern	✓
Make a model / act it out	✓
Try then improve	
Create a systematic list	✓
Solve a simpler but related problem	
Work backwards	

Key mathematical ideas

The aim of this task is for children to extend their experience of finding fractions of amounts by following and adapting a story. Poppy the penguin starts off with 24 fish, but gives certain fractions of her fish away to three other penguins: Perry, Padma and Pablo. The fraction given to each penguin is different ($\frac{1}{2}$, $\frac{1}{3}$ and $\frac{1}{4}$), and children can choose which fraction is given to each penguin. They must remember that after the first fraction is given away, the 'whole' is no longer 24; it is the number of fish that Poppy still has left.

Children will discover that, no matter what order the fractions are used, Poppy is always left with 6 fish. This is because the whole is always the same in each scenario (24 fish), as are the fractions. Because the denominators are all factors of 24, no matter the order they are used in, the end result will always be the same. However, the order does make a difference to the number of fish for the other three penguins.

Ensure that children are confident in the sharing method for division: where the quotient represents the number of shared fish in each group rather than the number of groups the total is being divided into. This will help children to be clear about what the numerical answer means.

For the Communicate task children are challenged to write their own fraction story. They should be encouraged to think of a different total to start with, ensuring it is a multiple of all the denominators they select for their fractions. Some children may choose to use non-unit fractions such as $\frac{2}{4}$, $\frac{2}{3}$ and $\frac{3}{4}$.

Answers are given in situ throughout the tasks because of the cumulative nature of the tasks.

Teaching guidance

If you are using the **Express Pathway**, please turn to pages 108–113.

Teaching guidance for the **Full 6C Pathway** is given below.

The **Teacher's Presentation** PowerPoint file has slides specifically mentioned in this teaching guidance. Some of the slides include notes (within the PowerPoint file) providing 'stumble support' suggestions, that may help you to support children in their problem-solving thinking and enable access for children who are less confident.

Awe and wonder

Reveal slides 1–2 one at a time, allowing time for discussion and observations as prompted below.

Slide 1

- I can see some sentences about a penguin and some fish.
- I have noticed that there are 24 fish.
- I can see one fraction in the sentences.

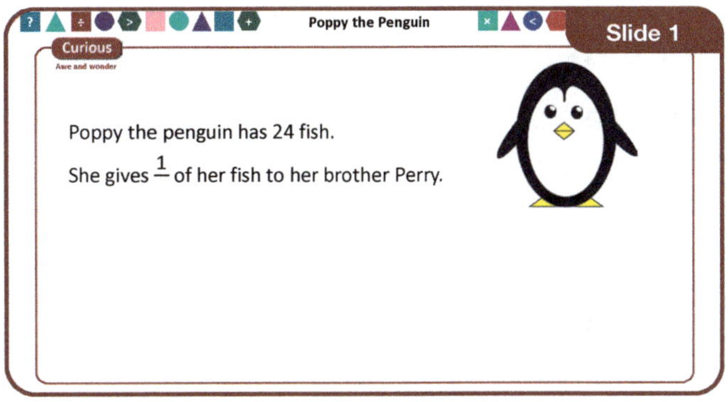

Slide 2

- I can see that there are some more fractions in the sentences. There are three fractions.
- I have noticed that the fractions aren't complete. There are missing numbers.
- I have noticed that the numerators for the fractions are all 1.
- It reminds me of when I have helped share the pencils out.

Now share with the children that the numbers missing from the fractions are 2, 4 and 3.

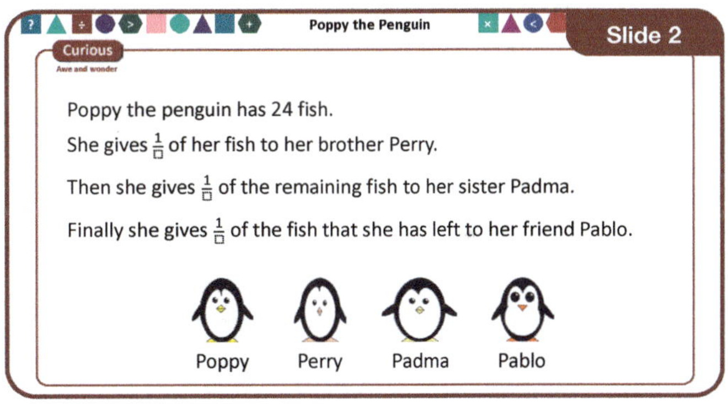

Develop and deepen

Slide 3

Display slide 3.

Encourage children to use the words in the word bank to develop and deepen their thoughts and questions.

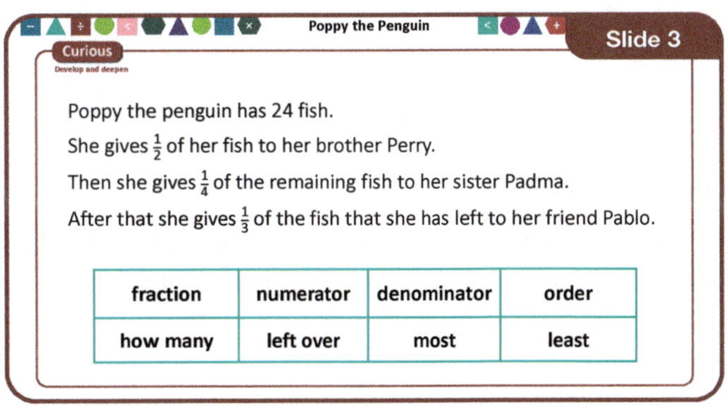

fraction	numerator	denominator	order
how many	left over	most	least

Possible lines of enquiry linked to the word bank:

- I wonder how many fish Poppy has at the end.
- I wonder who has the most / least fish at the end.
- I wonder which order the missing denominators go in.
- I wonder what changes and what stays the same if we change the order of the fractions.

Possible lines of enquiry that children may raise now, but should be held back until the **Create** task:

- I wonder what would happen if the numerators were not always 1.
- I wonder if I can write my own fraction story.

Consider whether / how to collate and display all the suggested questions, and whether / how to steer children towards selecting an appropriately challenging question.

Pupil Resources for Full and Express Pathways

Distribute the **Pupil Resources** for **Poppy the Penguin** to the children to help support them in their exploration of the different tasks. Use either of the **Curious Pupil Resources** depending on whether you are following the **Full Pathway** or the **Express Pathway**. The **Pupil Resources** for the remaining 5Cs are common to both pathways.

To provide additional flexibility in the classroom, each of the **Pupil Resources** is also provided in the **Teacher's Presentation** PowerPoint file (see slides 4–7), so you can display these at any time if you wish.

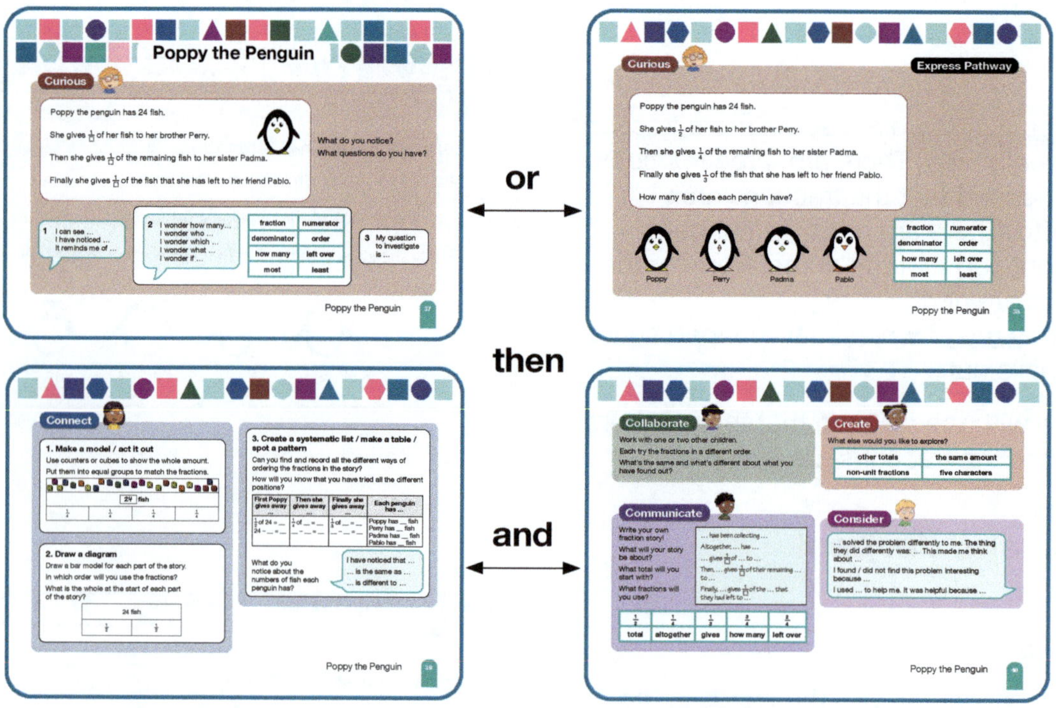

Express Pathway

If you are following the **Express Pathway**, spend a few minutes discussing the **Curious Pupil Resource** with the children. (This is slide 5, so you could display it if you wish.) Use some of the prompts on the previous page to help.

Three tasks involving different strategies are suggested for children to make connections.

1. Make a model / act it out

Additional resources required: counters or interlocking cubes

Slides 8, 9 and 10

Print slides 8, 9 and 10 and distribute to children.

Allow children to decide the order in which they use the three fractions. They should use their print-outs of the slides in their chosen order.

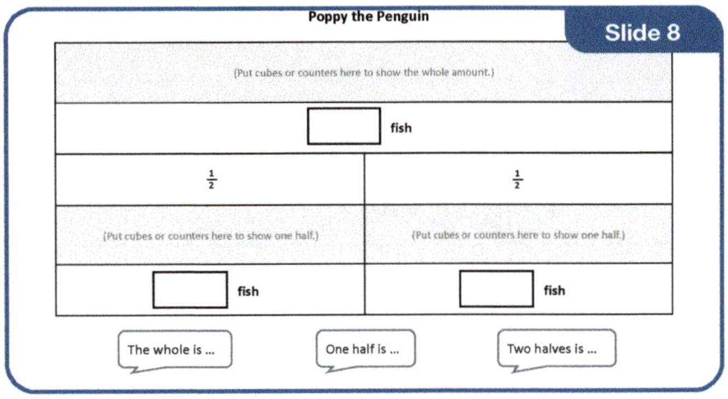

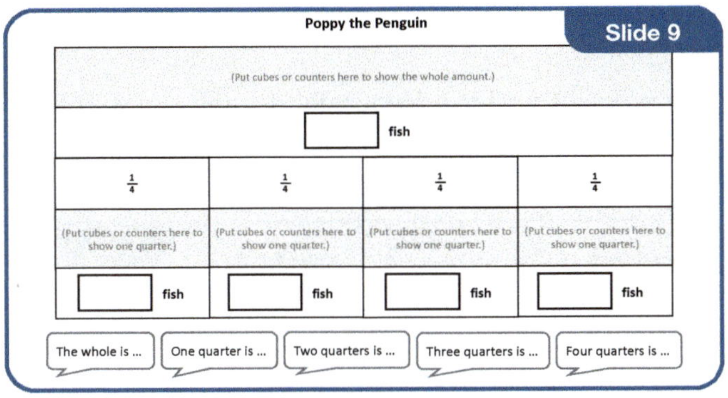

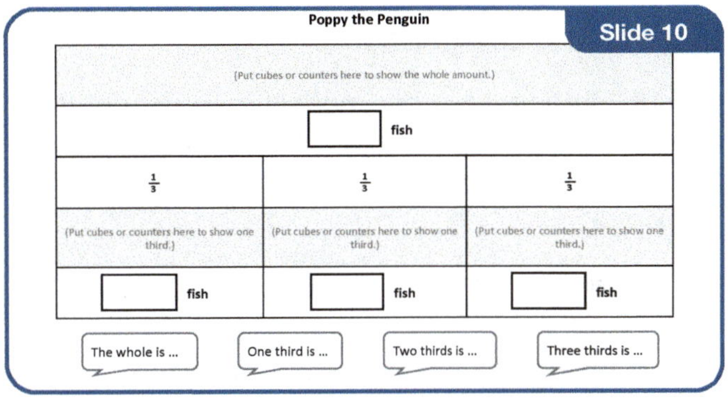

Ask children to choose a concrete resource (for example, counters or interlocking cubes). They count out the correct number to represent the whole amount, and place them in the top row of their first bar model. They can then share the resources between the empty boxes to show a half / quarter / third.

The following example is based on a child choosing to use the fractions in the order $\frac{1}{4}$, $\frac{1}{3}$, $\frac{1}{2}$.

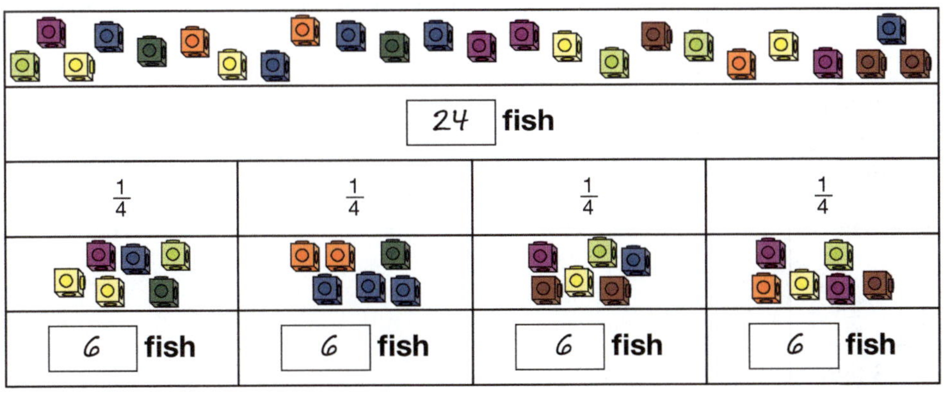

- Poppy gives a quarter of her fish to Perry. One-quarter of 24 is 6. 24 subtract 6 is 18, so Poppy has 18 fish left.
- *What happens next?*

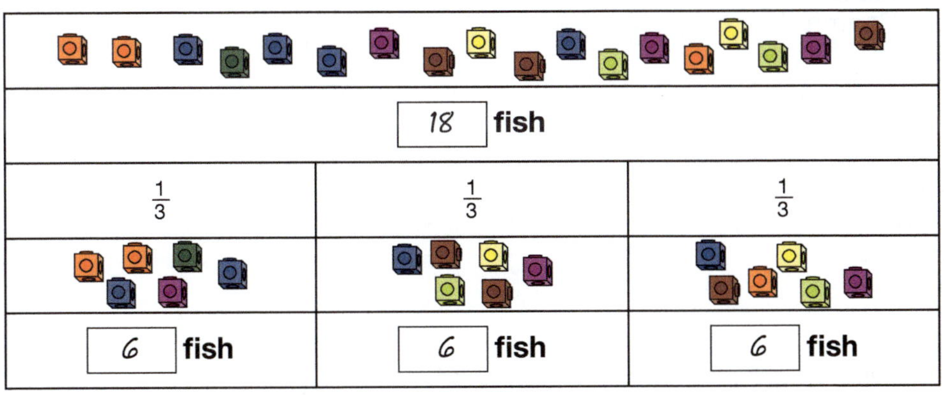

- The whole amount Poppy has at the start now is 18 fish.
- She gives one-third of the fish to Padma. One-third of 18 is 6. 18 subtract 6 is 12, so Poppy now has 12 fish left.
- *What happens next?*

12 fish	
$\frac{1}{2}$	$\frac{1}{2}$
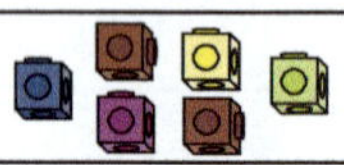	
6 fish	6 fish

- The whole amount Poppy has now is 12 fish.
- She gives one-half to Pablo. One-half of 12 is 6. 12 subtract 6 is 6, so Poppy has 6 fish left at the end.

2. Draw a diagram

Ask children to draw bar models to represent the whole amount and the fractions at each point in the story. It is important to make sure all children realise that the 'whole' at each point in the story changes, and needs to be taken into account before they find the next fraction of fish.

Slide 11

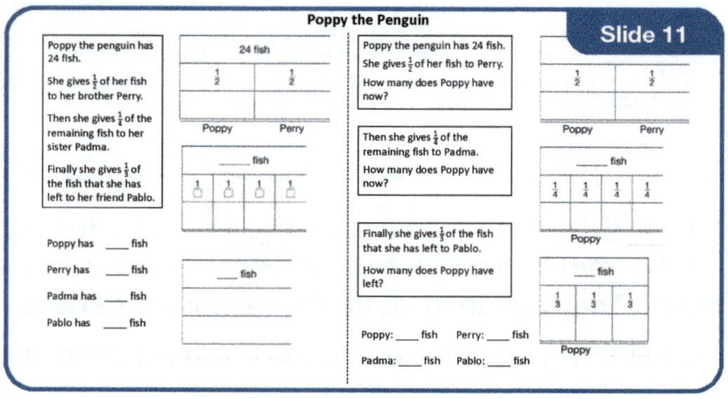

Print slide 11 and distribute to any children who you feel would benefit from some scaffolding.

The following example is based on using the fractions in the order $\frac{1}{2}$, $\frac{1}{4}$, $\frac{1}{3}$. This order could be used as a model and then children could explore what happens if they choose a different order.

24 fish	
$\frac{1}{2}$	$\frac{1}{2}$
12 fish	12 fish
Poppy	Perry

12 fish			
$\frac{1}{4}$	$\frac{1}{4}$	$\frac{1}{4}$	$\frac{1}{4}$
3 fish	3 fish	3 fish	3 fish
Poppy			Padma

9 fish		
$\frac{1}{3}$	$\frac{1}{3}$	$\frac{1}{3}$
3 fish	3 fish	3 fish
Poppy		Pablo

- Poppy has 6 fish. Perry has 12 fish. Padma has 3 fish. Pablo has 3 fish.

3. Create a systematic list / make a table / spot a pattern

Additional resources required: counters or interlocking cubes (optional)

Slide 12

Print slide 12 and distribute to children.

- *Can you find and record all the different ways of ordering the fractions in the story?*
- *How will you know that you have tried all the different positions?*
- *Do you think each penguin will get the same number of fish in each scenario?*

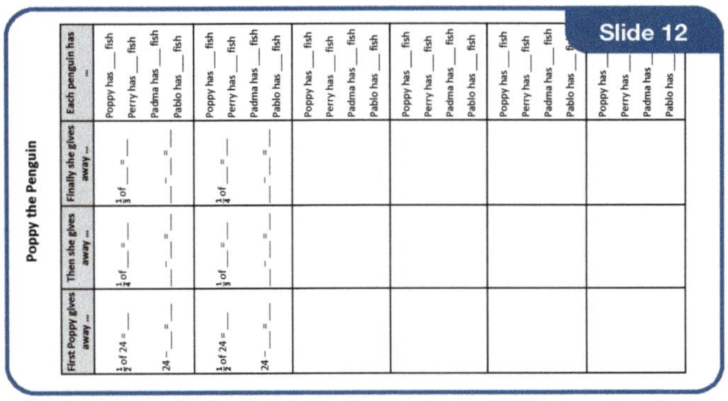

Slide 13

Display slide 13 to show the first two rows of the table. Model completing the first row, so children understand how to record the information in their own tables.

Children could also use print-outs of slides 8, 9 and 10 with concrete resources to model each stage of the story for each scenario.

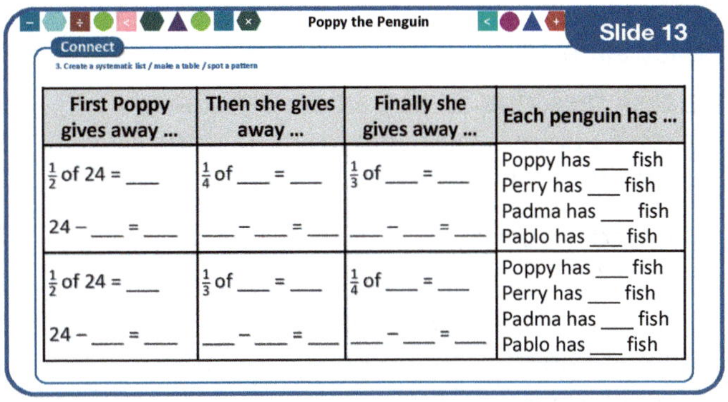

First Poppy gives away ...	Then she gives away ...	Finally she gives away ...	Each penguin has ...	
$\frac{1}{2}$ of 24 = 12 24 – 12 = 12	$\frac{1}{3}$ of 12 = 4 12 – 4 = 8	$\frac{1}{4}$ of 8 = 2 8 – 2 = 6	Poppy has 6 fish. Perry has 12 fish.	Padma has 4 fish. Pablo has 2 fish.
$\frac{1}{2}$ of 24 = 12 24 – 12 = 12	$\frac{1}{4}$ of 12 = 3 12 – 3 = 9	$\frac{1}{3}$ of 9 = 3 9 – 3 = 6	Poppy has 6 fish. Perry has 12 fish.	Padma has 3 fish. Pablo has 3 fish.
$\frac{1}{3}$ of 24 = 8 24 – 8 = 16	$\frac{1}{2}$ of 16 = 8 16 – 8 = 8	$\frac{1}{4}$ of 8 = 2 8 – 2 = 6	Poppy has 6 fish. Perry has 8 fish.	Padma has 8 fish. Pablo has 2 fish.
$\frac{1}{3}$ of 24 = 8 24 – 8 = 16	$\frac{1}{4}$ of 16 = 4 16 – 4 = 12	$\frac{1}{2}$ of 12 = 6 12 – 6 = 6	Poppy has 6 fish. Perry has 8 fish.	Padma has 4 fish. Pablo has 6 fish.
$\frac{1}{4}$ of 24 = 6 24 – 6 = 18	$\frac{1}{2}$ of 18 = 9 18 – 9 = 9	$\frac{1}{3}$ of 9 = 3 9 – 3 = 6	Poppy has 6 fish. Perry has 6 fish.	Padma has 9 fish. Pablo has 3 fish.
$\frac{1}{4}$ of 24 = 6 24 – 6 = 18	$\frac{1}{3}$ of 18 = 6 18 – 6 = 12	$\frac{1}{2}$ of 12 = 6 12 – 6 = 6	Poppy has 6 fish. Perry has 6 fish.	Padma has 6 fish. Pablo has 6 fish.

- *What patterns can you see in your table?*
- I have noticed that Poppy always has 6 fish left!
- I have noticed that the other penguins don't always have the same amount. This changes when I change the order of the fractions.
- If I use a half first, Perry gets more fish than anyone else because we are only sharing by 2.
- If you do a quarter, then a third, then a half, all the penguins end up with the same number of fish!

Collaborate

Refer to the **Collaborate** task of the **Pupil Resource**.

This task provides an opportunity for children to work in pairs or threes, with each child exploring a different order of the fractions.

For example:

- one child exploring using half first and then alternating the quarter and third;
- one child exploring using a quarter first;
- one child exploring using a third first.

The children could then join back together as a group to compare their findings.

Communicate

Refer to the **Communicate** task of the **Pupil Resource**.

Slide 14

Display slide 14.

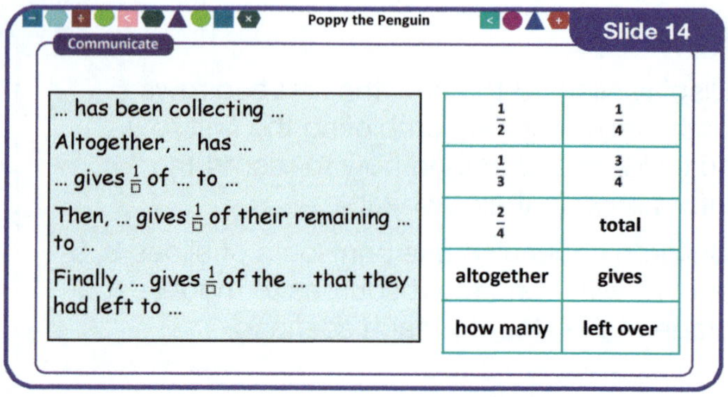

This task provides an opportunity for the children to create their own fraction story. They will need to think about:

- the fractions they use (this could also include $\frac{3}{4}$ and $\frac{2}{4}$);
- the total that they will start with;
- they will need to test their story to make sure that it works;
- whether their character will keep the larger amount (as Poppy does) or give away the larger amount.

The writing frame from slide 14 could be used to give support.

Pairs could then present their story as a poster which could be shared and explored by another pair.

Create

Refer to the **Create** task of the **Pupil Resource**.

- *Were there any questions asked at the start of the task that you now want to think about?*
- *Is there anything new that you are wondering now?*

Children can use the word bank to prompt further questions of their own.

other totals	**the same amount**
non-unit fractions	**five characters**

- I wonder what other totals would work for the story about Poppy the penguin.
- I wonder if I could write a fraction story where all the characters end up with the same amount.
- I wonder what would happen if the numerators were not always 1.
- I wonder if I can write my own fraction story.

Refer to the **Consider** task of the **Pupil Resource**.

Slide 15

Display slide 15.

Encourage children to think about the strategies that they have used. The prompts below may help get children started.

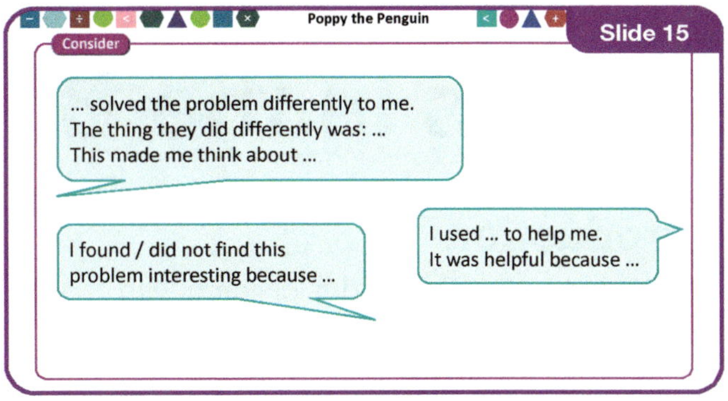

Cognitive learning domain

- The trickiest part of solving the problem was …
- … solved the problem differently to me. The thing they did differently was: … This made me think about …
- I am sure / not sure about my answer because …

Affective domain

- I found / did not find this problem interesting because …
- I felt … when I found an answer because …
- I was frustrated by … because …

Metacognitive domain

- I got stuck when … I unstuck myself by …
- I used … to help me. It was helpful because …
- Next time, I will …

What's in the Box?

Year 1 Year 2

Introduction

What's in the Box? provides children with an opportunity to rehearse different coin combinations and apply their addition skills. In order to 'win' the boxes, children need to figure out the combination of coins inside each one. Curiosity can be increased by recreating the mystery boxes in the classroom and only allowing children to open them once they have found the correct combination.

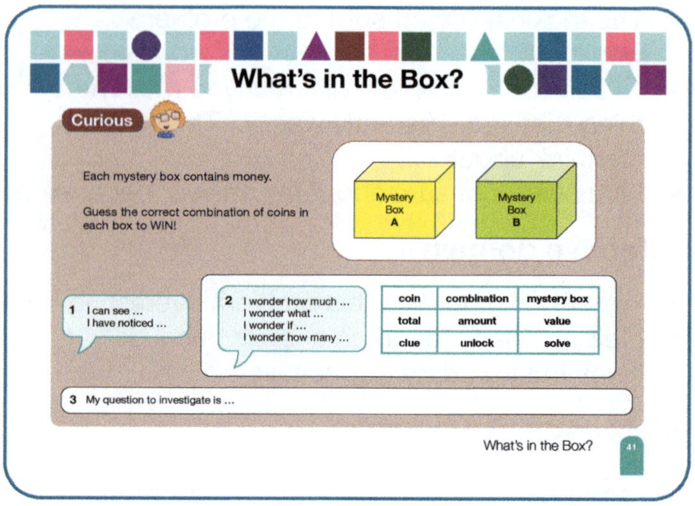

There are three clues for each box, structured to increasingly narrow down the number of combinations possible. Clues could all be provided at once or suspense can be built by providing the clues one at a time.

Mystery Box A is most appropriate for Year 1.

Mystery Box B is most appropriate for Year 2, although Year 2 children may benefit from solving Mystery Box A first.

This task promotes greater-depth learning in a range of ways, supporting children to identify and articulate underlying mathematical structures. For example, in the **Connect** stage, greater depth is promoted by supporting children to choose an appropriate strategy and explain their choice, and understand that sometimes more than one answer is possible.

This **Teacher Guide** is supported by photocopiable **Pupil Resources** and **Teacher's Presentation** PowerPoint slides for whole-class or group display.

National Curriculum

What's in the Box? covers these Year 1 and Year 2 National Curriculum objectives.

Year 1 Addition and subtraction

- Add and subtract one-digit and two-digit numbers to 20, including zero.

Year 1 Measurement

- Recognise and know the value of different denominations of coins and notes.

Year 2 Addition and subtraction

- Add and subtract numbers using concrete objects, pictorial representations, and mentally, including: a two-digit number and ones; a two-digit number and tens; two two-digit numbers; adding three one-digit numbers.

Year 2 Measurement

- Find different combinations of coins that equal the same amounts of money.
- Solve simple problems in a practical context involving addition and subtraction of money of the same unit, including giving change.

The 6C cycle

The grid on the right shows the relative weighting given to each of the 6 Cs in **What's in the Box?**

Curious	★★
Connect	★★★
Collaborate	★★
Communicate	★★
Create	★★
Consider	★

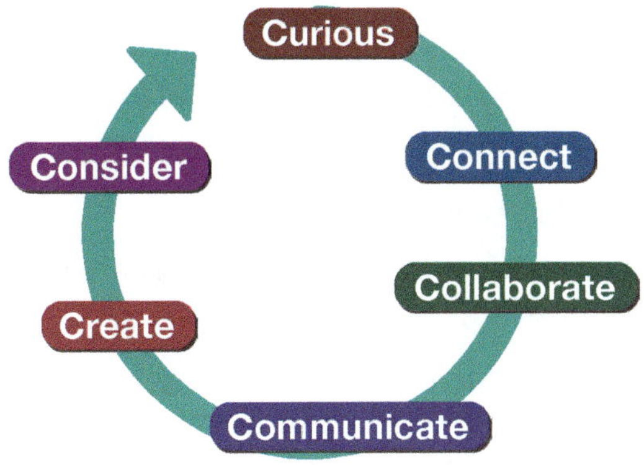

Problem-solving heuristics

The grid shows which problem-solving strategies children are likely to use when investigating the **Connect** tasks in **What's in the Box?** These strategies are further exemplified in the **Connect** stage below (pages 118–120).

Draw a diagram	✓
Make a table / spot a pattern	
Make a model / act it out	✓
Try then improve	
Create a systematic list	
Solve a simpler but related problem	✓
Work backwards	

Key mathematical ideas

The aim of this task is for children to practise coin recognition and mental addition strategies with small numbers.

Useful strategies could include looking for complements to 10 (for example, 2p + 5p + 5p could be rearranged to 5p + 5p + 2p).

This task is an opportunity for children to explore all possibilities and develop systematic management of their answers. However, it is not necessary for children to have found all possibilities of coin combinations to solve the clues.

It is important that children have access to coins to support them throughout. However, money is challenging for younger children as it is not proportional (for example, 2p coins are larger than 5p coins, but are worth less). In order to make calculations easier, coins could be attached to concrete resources (for example, number rods or trains of interlocking cubes) to make proportional models.

Answers to the tasks are given on page 122.

Teaching guidance

If you are using the **Express Pathway**, please turn to pages 118–122.

Teaching guidance for the **Full 6C Pathway** is given below.

The **Teacher's Presentation** PowerPoint file has slides specifically mentioned in this teaching guidance. Some of the slides include notes (within the PowerPoint file) providing 'stumble support' suggestions, that may help you to support children in their problem-solving thinking and enable access for children who are less confident.

Awe and wonder

If possible, recreate the boxes in the classroom to increase the level of curiosity and engagement.

Slide 1

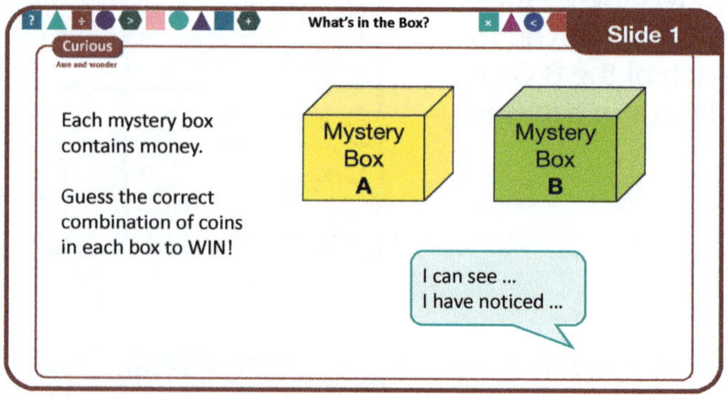

Display slide 1.

Ask children to share what they are noticing or wondering.

- I can see two mystery boxes.
- I have noticed that the boxes contain money.
- I have noticed that the boxes only contain coins, and no notes.
- I have noticed that to win the box you need to work out the right combination of coins.

Now explain that, to open each box, it is not enough to find any combination of coins that adds up to the mystery total. They have to find a specific combination.

- *Before I tell you any more about the mystery boxes, let's practise finding coin combinations.*
- *What if one of the totals was 6p? What are the different ways we could make this total?*

Together, record all the possibilities:

5p + 1p　　2p + 2p + 2p　　2p + 2p + 1p + 1p

2p + 1p + 1p + 1p + 1p　　1p + 1p + 1p + 1p + 1p + 1p

- *What if the next clue was: 'all the coins are the same'? Which of our possibilities could be the right combination?*

Guide children through the process of narrowing down the list to two possibilities:

2p + 2p +2p　　1p + 1p + 1p + 1p + 1p + 1p

- *What if the third clue was: 'there are more than three coins in the box'?*

Agree that the only combination that fits all three clues is 1p + 1p + 1p + 1p + 1p + 1p.

Develop and deepen

Slide 2

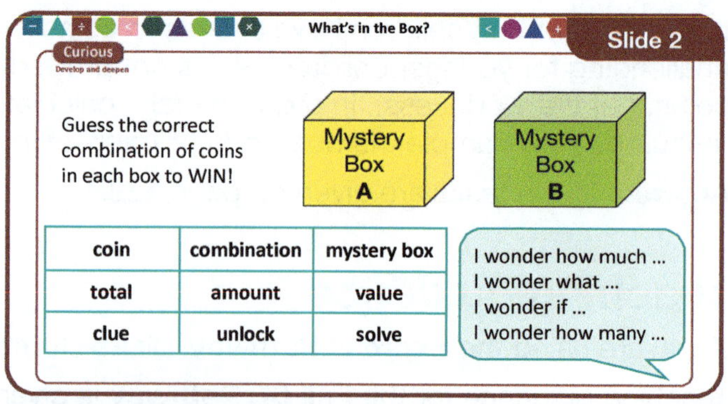

Display slide 2.

Encourage children to use the words in the word bank to develop and deepen their thoughts and questions.

coin	combination	mystery box
total	amount	value
clue	unlock	solve

Possible lines of enquiry:

- I wonder how much money is in each box.

Possible lines of enquiry linked to the word bank:

- I wonder what the clues might be.
- I wonder if there will be more than one right answer.
- I wonder how many clues it takes to get to one answer.
- I wonder if I can work out the coin combinations to unlock the mystery boxes.

Possible lines of enquiry that children may raise now, but should be held back until the **Communicate** and **Create** tasks:

- I wonder if I could make my own 'mystery box' puzzle.
- I wonder if I could write a 'mystery box' puzzle that needs more clues to be able to solve it.
- I wonder if I could write a 'mystery box' puzzle that includes notes.

Consider whether / how to collate and display all the suggested questions, and whether / how to steer children towards selecting an appropriately challenging question.

Slides 3 and 4

Display slide 3 or 4, depending on which mystery box you want children to focus on.

You may wish to reveal the clues one at a time, by covering the parts of the slide you do not yet want children to see. Or you may wish to display / reveal the clues in a different way.

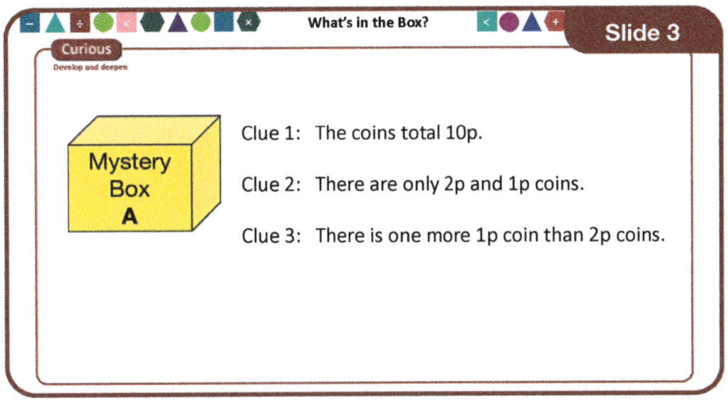

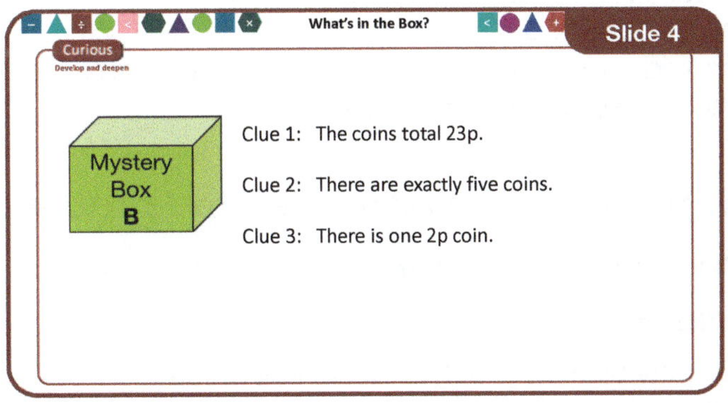

Pupil Resources for Full and Express Pathways

Distribute the **Pupil Resources** for **What's in the Box?** to the children to help support them in their exploration of the different tasks. Use either of the **Curious Pupil Resources** depending on whether you are following the **Full Pathway** or the **Express Pathway**. The **Pupil Resources** for the remaining 5Cs are common to both pathways.

To provide additional flexibility in the classroom, each of the **Pupil Resources** is also provided in the **Teacher's Presentation** PowerPoint file (see slides 5–8), so you can display these at any time if you wish.

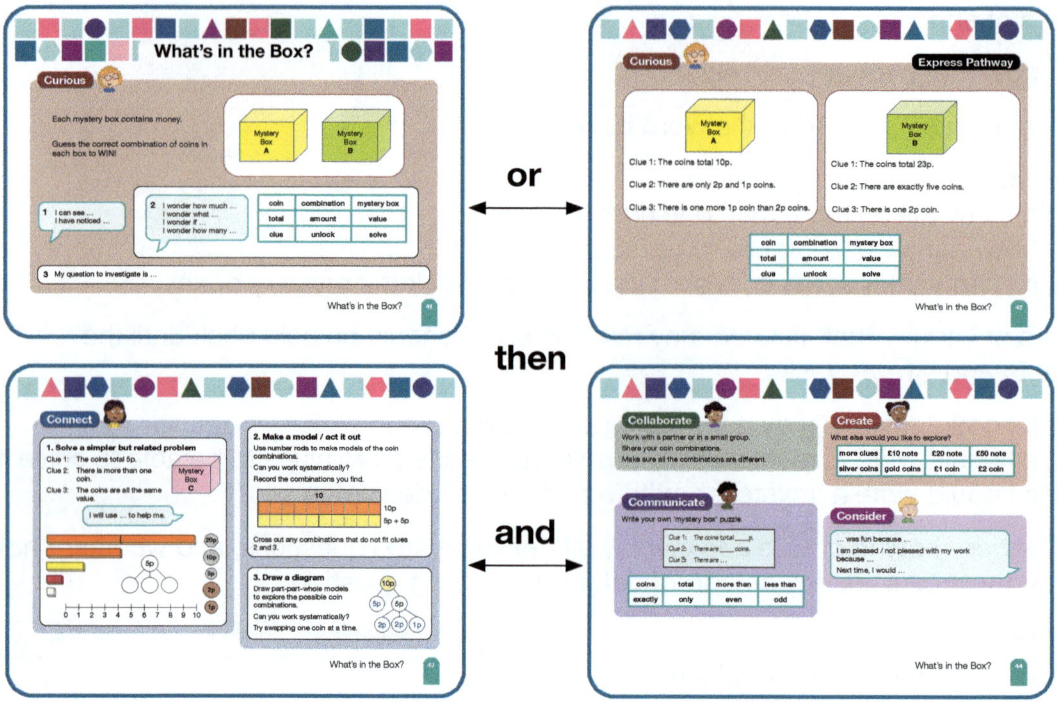

Express Pathway

If you are following the **Express Pathway**, spend a few minutes discussing the **Curious Pupil Resource** with the children. (This is slide 6, so you could display it if you wish.) Use some of the prompts on the previous page to help.

Three tasks involving different strategies are suggested for children to make connections.

Please note that, for children working on Mystery Box B, it should not be expected that they will find all 54 possibilities for clue 1 by themselves. When they have built up a good bank of possibilities, they should move on to clue 2. They can then combine the information from the first two clues to target their exploration more closely.

1. Solve a simpler but related problem

Additional resources required: concrete resources (for example, coins, number rods, bead strings)

Slide 9

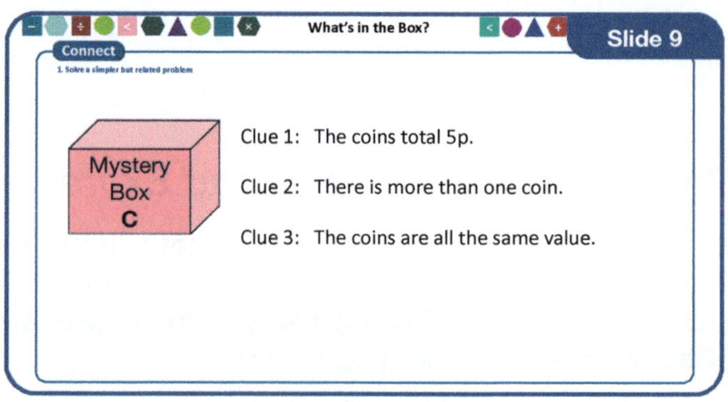

Display slide 9, which shows three clues for Mystery Box C.

This simpler example will give children an opportunity to practise working systematically.

Work through the clues, asking children to record all the possibilities each time. They could do this by making models, by drawing diagrams (for example, part-part-whole models or number lines), or just by jotting down notes.

Clue 1: The coins total 5p.

Clue 2: There is more than one coin.

Clue 3: The coins are all the same value.

2. Make a model / act it out

Additional resources required: number rods, coloured pencils to match the number rods used

Slides 10 and 11

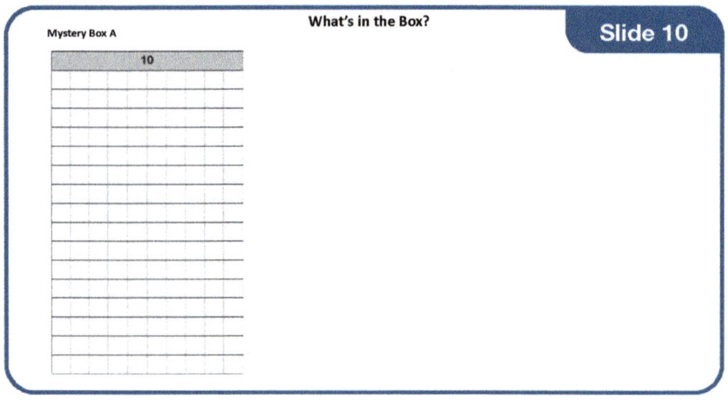

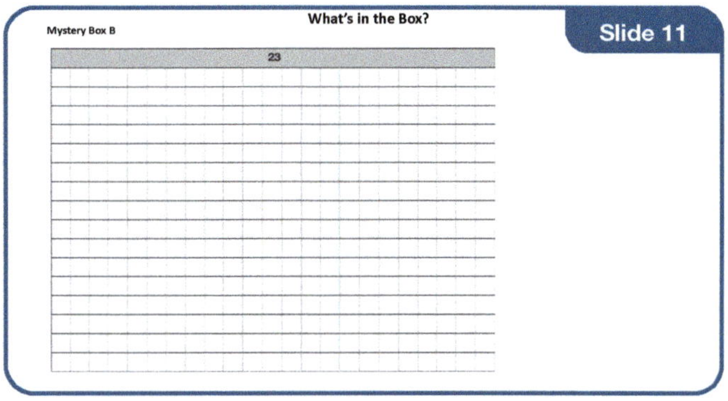

Print slide 10 or 11 and distribute to children (depending whether they are working on Mystery Box A or Mystery Box B).

If working on Mystery Box B, children are likely to need more than one copy of slide 11.

When printed, the recording grids should be the correct size for number rods.

Children can lay number rods along the rows to find combinations that total the given value.

Children will probably work randomly at first. Support them to rearrange their models to make patterns easier to see.

- *Can you see any combinations that are missing from the pattern?*

Children can record the combinations by colouring the parts to match the number rods, and writing the matching calculation for each row.

10p
5p + 5p
5p + 2p + 2p + 1p
5p + 2p + 1p + 1p + 1p
5p + 1p + 1p + 1p + 1p + 1p
2p + 2p + 2p + 2p + 2p
2p + 2p + 2p + 2p + 1p + 1p
2p + 2p + 2p + 1p + 1p + 1p + 1p
2p + 2p + 1p + 1p + 1p + 1p + 1p + 1p
2p + 1p + 1p + 1p + 1p + 1p + 1p + 1p + 1p
1p + 1p + 1p + 1p + 1p + 1p + 1p + 1p + 1p + 1p

Completed recording grid for Mystery Box A, showing all 11 possibilities after clue 1

- *For Mystery Box B, how can we represent 20p?* (Tape two 10-rods together.)
- *How can we make sure we remember that this means 20p, and not two 10p coins?* (When we record it on the sheet we could write '20p'.)

For clues 2 and 3, children could draw a cross next to any coin combinations that no longer fit.

3. Draw a diagram

Some children may prefer to draw diagrams rather than making concrete models. For example, they could draw part-part-whole models to explore the different ways to make the given total. They should still have access to coins and concrete resources to help them if they would like.

Slides 12 and 13

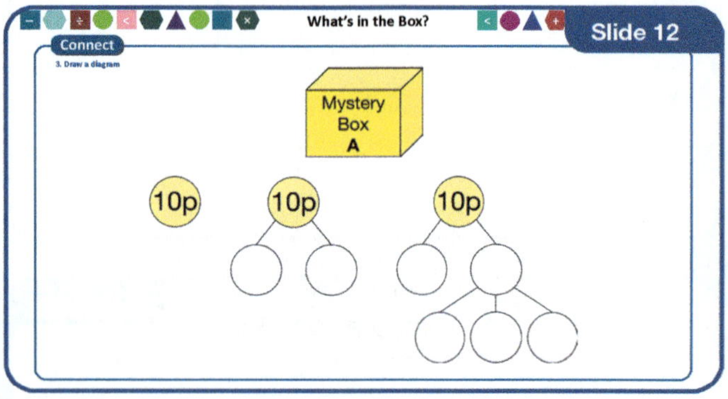

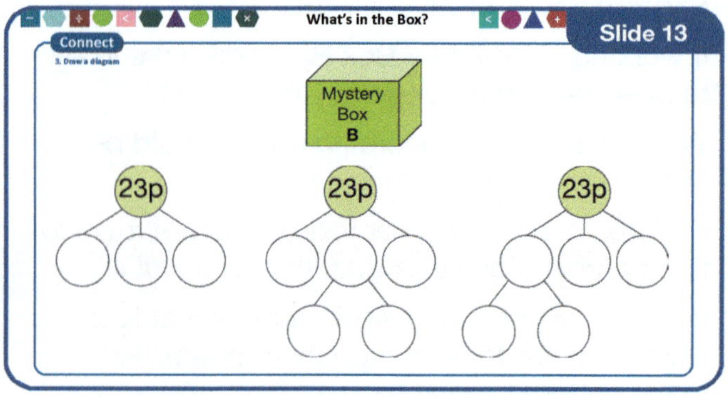

Display slide 12 or 13 (depending whether children are working on Mystery Box A or Mystery Box B).

Ask children to use the blank models on the slide as a starting point for their own. Encourage them to work systematically by thinking about which coin they can swap out and replace with other coins, but only changing one coin at a time.

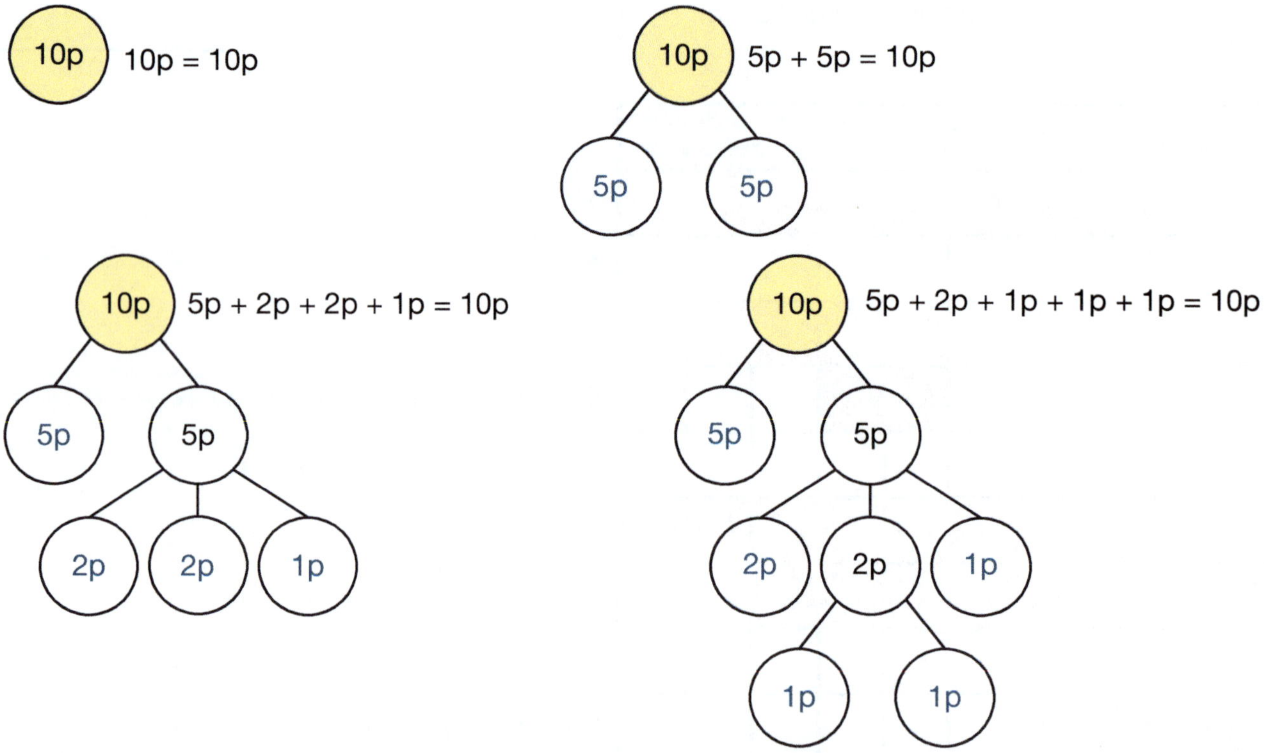

Four part-part-whole models showing different coin combinations for Mystery Box A after clue 1

For clues 2 and 3, children could cross out any coin combinations that no longer fit.

Collaborate

Refer to the **Collaborate** task of the **Pupil Resource**.

This task provides an opportunity for children to share their calculations that satisfy clue 1 so they have a larger data set. This could be done in pairs or small groups.

For Mystery Box A, the focus could be on checking for duplicate combinations and agreeing on a single set of unique combinations (there are 11 in total). For example, if one child has recorded 5p + 2p + 2p + 1p and another has recorded 5p + 1p + 2p + 2p:

- *Do these two calculations represent two different sets of coins?*

If needed, ask children to match actual coins up with each of the two calculations to check.

For Mystery Box B, the focus could be on pooling all the combinations that children have come up with, to try to get closer to a complete set (there are 54 in total, so children are unlikely to have found them all by themselves). Encourage children to focus on finding a way to order the possibilities systematically to check whether any are missing.

Communicate

Refer to the **Communicate** task of the **Pupil Resource**.

Slide 14

Display slide 14.

This task provides an opportunity for children to create their own mystery box puzzle.

- I wonder if I could make my own 'mystery box' puzzle.

They can use the sentence frames and word bank to help them.

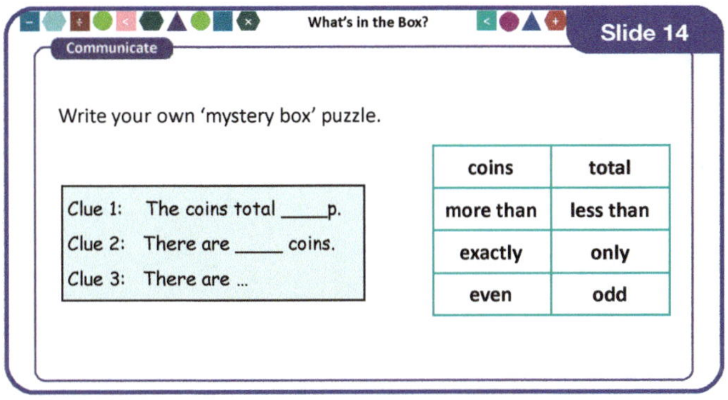

coins	total	more than	less than
exactly	only	even	odd

Create

Refer to the **Create** task of the **Pupil Resource**.

Slide 15

Display slide 15.

- *Were there any questions asked at the start of the task that you now want to think about?*
- *Is there anything new that you are wondering now?*

Children can use the word bank to prompt further questions of their own.

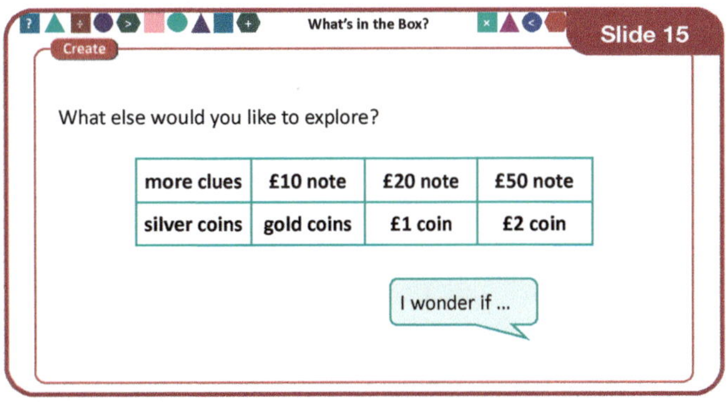

more clues	£10 note	£20 note	£50 note
silver coins	gold coins	£1 coin	£2 coin

- I wonder if I could write a 'mystery box' puzzle that needs more clues to be able to solve it.
- I wonder if I could write a 'mystery box' puzzle that includes notes.

Consider

Refer to the Consider task of the **Pupil Resource**.

Encourage children to think about the strategies that they have used. The prompts below may help get children started.

Cognitive learning domain

- The trickiest part of solving the problem was …
- The new strategies that I learned were …

Affective domain

- … was fun because …
- I am pleased / not pleased with my work because …

Metacognitive domain

- I used … to help me. It was helpful because …
- Next time, I would …
- I would give these top tips about working well in a group: …

Answers

Mystery Box A

2p + 2p + 2p + 1p + 1p + 1p + 1p

Mystery Box B

10p + 5p + 5p + 2p + 1p

Mystery Box C

1p + 1p + 1p + 1p + 1p

Built to Scale

Year 2

Introduction

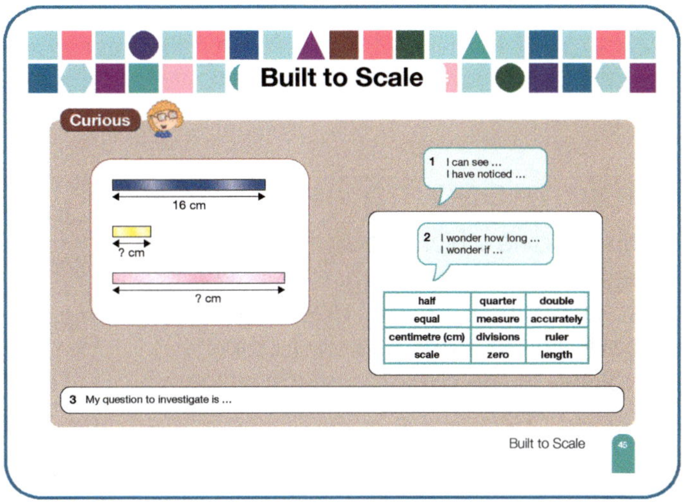

Built to Scale provides children with an opportunity to explore the proportional relationship that is present in all scales. They will start by creating their own ruler given one measurement. They will explore how a measurement point can be accurately discovered though halving, quartering and adding to known points. This strengthens their understanding of how scales are made and prepares them for scales with missing intervals.

Children will then have the opportunity to apply this learning in pairs to create their own unit of measurement. They will create a measuring tool to help others use their new unit. Discussions will focus on what this unit would be appropriate for measuring.

During the **Create** task children can take their exploration further and consider creating new units of measure for capacity, mass or time.

This task promotes greater-depth learning in a range of ways, supporting children to identify and articulate underlying mathematical structures. For example, in the **Create** stage, greater depth is promoted by supporting children to form their own questions that are creative and inventive.

This **Teacher Guide** is supported by photocopiable **Pupil Resources** and **Teacher's Presentation** PowerPoint slides for whole-class or group display.

National Curriculum

Built to Scale covers these Year 2 National Curriculum objectives.

Measurement

- Choose and use appropriate standard units to estimate and measure length / height in any direction (m / cm); mass (kg / g); temperature (°C); capacity (litres / ml); to the nearest appropriate unit, using rulers, scales, thermometers and measuring vessels.

Fractions

- Recognise, find, name and write fractions $\frac{1}{3}$, $\frac{1}{4}$, $\frac{2}{4}$ and $\frac{3}{4}$ of a length, shape, set of objects or quantity.
- Write simple fractions for example, $\frac{1}{2}$ of 6 = 3 and recognise the equivalence of $\frac{2}{4}$ and $\frac{1}{2}$.

The 6C cycle

The grid on the right shows the relative weighting given to each of the 6 Cs in **Built to Scale**.

Curious	★★
Connect	★★
Collaborate	★★
Communicate	★★
Create	★★★
Consider	★

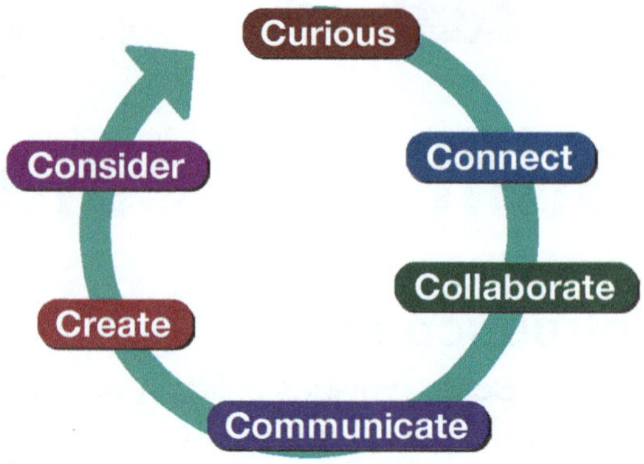

Problem-solving heuristics

The grid shows which problem-solving strategies children are likely to use when investigating the **Connect** tasks in **Built to Scale**. These strategies are further exemplified in the **Connect** stage below (pages 127–128).

Draw a diagram	
Make a table / spot a pattern	
Make a model / act it out	✓
Try then improve	
Create a systematic list	
Solve a simpler but related problem	
Work backwards	

Key mathematical ideas

In **Built to Scale**, children have to apply their knowledge of number magnitude, as well as halving and quartering. By building their own measuring tool, children will gain a deeper understanding of the size of a centimetre, enabling them to make more informed estimates before accurately measuring.

Creating their own measuring tool shows children that, by applying a simple pattern, they can use one fact to gain many others. 'If I know this, then I also know …'

The children are still in the early stages of understanding halves and quarters, so it is important that they continue to have hands-on experience of fractions as equal groups. Folding paper strips reinforces that halving is dividing into two equal groups, and that quarters can be found by dividing one half into two equal groups. They will see that two quarters are equal to one half.

Children will be provided with a blank strip of paper, at least 30 cm long, and a 16 cm length of ribbon (or string or paper). This is all they need to create their own accurate 30 cm ruler. They will mark 16 cm then fold the paper strip repeatedly to find and mark 8 cm, 4 cm, 2 cm and 1 cm. They will then fold back over to the other side of the 16 cm mark to find 24 cm (16 + 8). It is now possible to complete the ruler either by continuing to apply the halving method between 4 cm and 8 cm and so on; or by folding the newly found measurements on to the ruler (for example, 10 cm could be marked by folding 2 cm on to 8 cm).

The length of the shorter ribbon can be found by noticing that if you fold the 16 cm length ribbon in half and half again it is the same length as the shorter ribbon. The children could then calculate the length in centimetres by either halving 16 cm then halving again, or finding a quarter of 16 cm.

It is more challenging to find the length of the longer ribbon, and creating a ruler is helpful. Children can observe that it is only slightly longer than the 16 cm ribbon, but they need to prove the length rather than guessing. One option is to halve the shorter 4 cm long ribbon, then add this to the 16 cm long ribbon to equal the longest ribbon.

Answers to the tasks are given on page 130.

Teaching guidance

If you are using the **Express Pathway**, please turn to pages 126–130.

Teaching guidance for the **Full 6C Pathway** is given below.

The **Teacher's Presentation** PowerPoint file has slides specifically mentioned in this teaching guidance. Some of the slides include notes (within the PowerPoint file) providing 'stumble support' suggestions, that may help you to support children in their problem-solving thinking and enable access for children who are less confident.

Additional resources required: a set of three ribbons for each pair of children (4 cm, 16 cm and 18 cm long), strips of paper at least 30 cm long

Awe and wonder

Slide 1

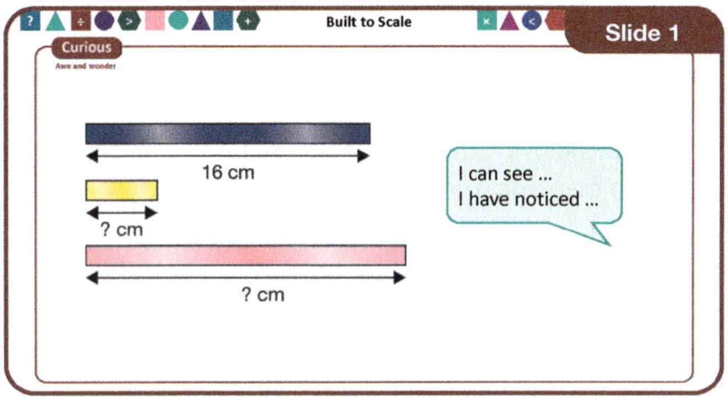

Display slide 1.

Provide each pair of children with a set of ribbons. Do not reveal the length of any ribbons.

⚠ Ensure all rulers have been removed from the classroom.

- <u>I can see</u> that the ribbons are different lengths.
- <u>I have noticed</u> that this ribbon is the longest.
- <u>I can see</u> that this ribbon is only a bit longer than this one.
- <u>I can see</u> that we all have the same ribbons.
- <u>I have noticed</u> that this one is between these two in length.
- <u>I have noticed</u> that one ribbon is about a quarter of the length of another one.

⚠ Watch out for children who are not using accurate language, for example, 'bigger' rather than 'larger', and 'longer' / 'shorter' rather than 'longest' / 'shortest' when comparing three items.

Develop and deepen

Slide 2

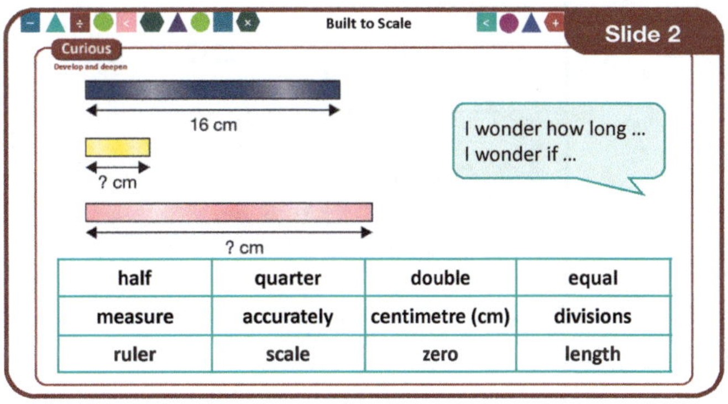

Display slide 2.

- *If we know that one ribbon is 16 centimetres long, how can we work out the lengths of the other two ribbons?*
- *You are not allowed to use a ruler!*

Provide each pair of children with a strip of paper which is at least 30 cm long.

Encourage children to use the words in the word bank to develop and deepen their thoughts and questions.

half	**quarter**	**double**	**equal**
measure	**accurately**	**centimetre (cm)**	**divisions**
ruler	**scale**	**zero**	**length**

Possible lines of enquiry:

- I wonder how long the other ribbons are.

Possible lines of enquiry linked to the word bank:

- I wonder if I can make an accurate ruler to measure in centimetres.
- I wonder if I can find the length of the other ribbons without making a ruler.

Possible lines of enquiry that children may raise now, but should be held back until the **Collaborate** task:

- I wonder if I can make a ruler using my own scale.

Consider whether / how to collate and display all the suggested questions, and whether / how to steer children towards selecting an appropriately challenging question.

Pupil Resources for Full and Express Pathways

Distribute the **Pupil Resources** for **Built to Scale** to the children to help support them in their exploration of the different tasks. Use either of the **Curious Pupil Resources** depending on whether you are following the **Full Pathway** or the **Express Pathway**. The **Pupil Resources** for the remaining 5Cs are common to both pathways.

To provide additional flexibility in the classroom, each of the **Pupil Resources** is also provided in the **Teacher's Presentation** PowerPoint file (see slides 3–6), so you can display these at any time if you wish.

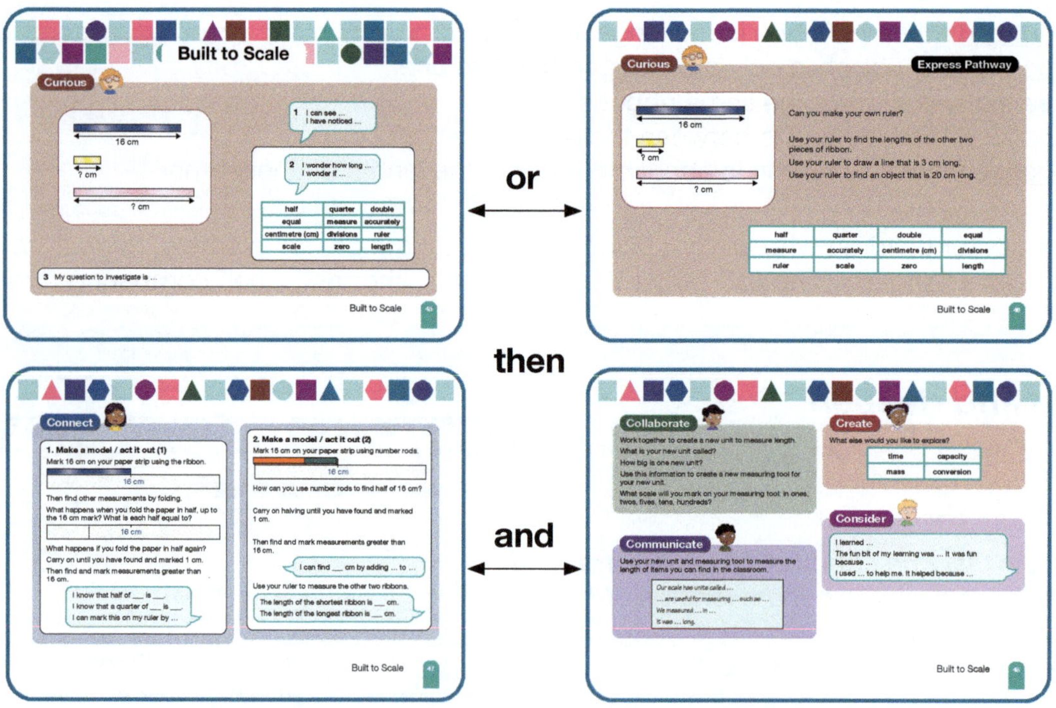

Express Pathway

If you are following the **Express Pathway**, spend a few minutes discussing the **Curious Pupil Resource** with the children. (This is slide 4, so you could display it if you wish.) Use some of the prompts on the previous page to help.

Two tasks involving different strategies are suggested for children to make connections.

1. Make a model / act it out (1)

Additional resources required: counters (optional)

Children could build their own ruler using their paper strip and their 16 cm ribbon as a reference point. Through accurate, sequential halving of the paper strip, they can identify 8 cm, 4 cm, 2 cm and 1 cm.

Slide 7

Display slide 7. This shows step 1 in the process: lining the end of the 16 cm ribbon up precisely with the end of the paper strip and making a mark where the ribbon ends.

- *What will you label the mark?* (16 cm)

⚠ Look out for children who do not line up the end of the ribbon with the end of the paper strip accurately.

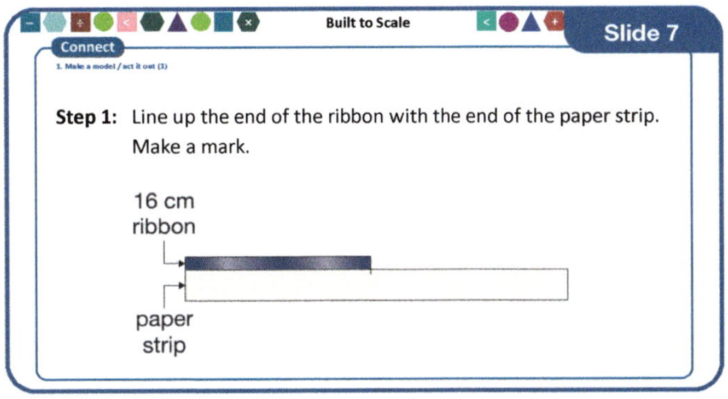

Slide 8

Display slide 8. This shows step 2 in the process: folding the paper strip so that the 'zero' end meets the 16 cm mark precisely.

- *What will you label the fold line?* (8 cm)
- *What will you do next?*

Support children to continue, folding three more times to find and mark 4 cm, 2 cm and 1 cm.

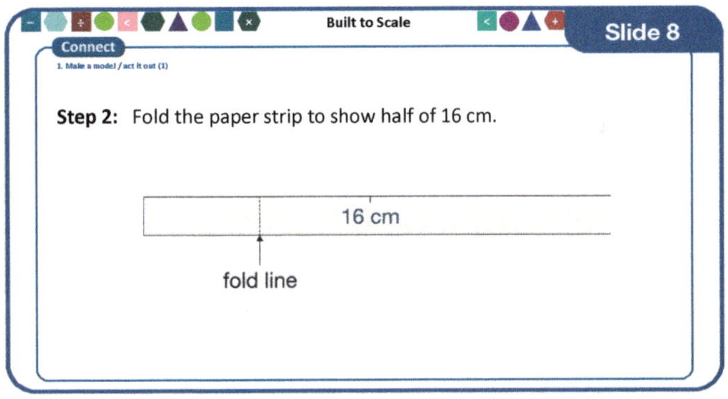

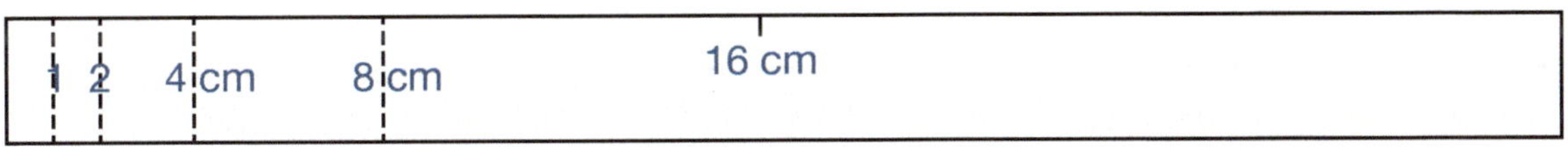

Then ask children to mark measurements greater than 16 cm on their paper strip. For example:

- *How could you find 24 cm?*
- *What is the difference between 24 cm and 16 cm?* (8 cm)
- I could fold my strip at the 16 cm mark, then mark where the 8 cm mark comes up to.

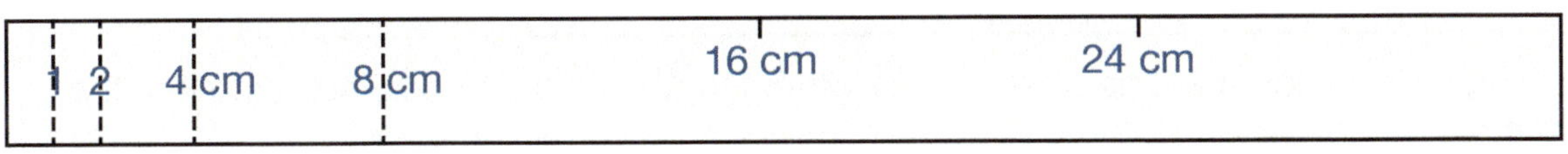

Children could continue marking more measurements until they have created a complete 30 cm ruler.

Slide 9

Print slide 9 if needed. This could be used to support any children who are not confident with finding halves and quarters of numbers. Use the grids to model how to share counters equally into groups.

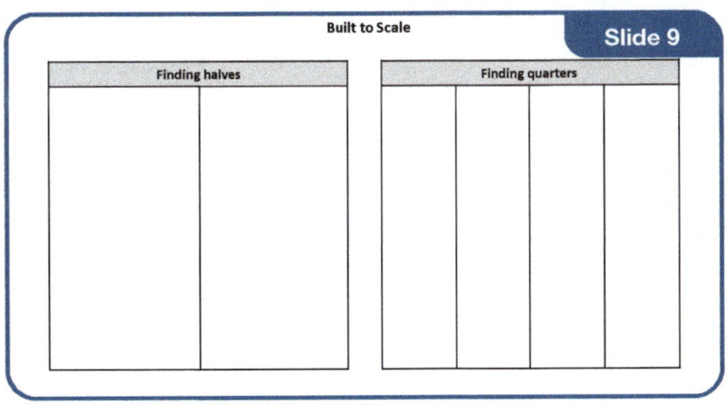

2. Make a model / act it out (2)

Additional resources required: number rods

As an alternative model method, children could use a concrete resource such as number rods (which are exact numbers of centimetres), especially if they are struggling to accurately fold the paper strip or to calculate halves.

Slide 10

Display slide 10.

Ask children to use a 10-rod and a 6-rod to find and mark 16 cm on their paper strip.

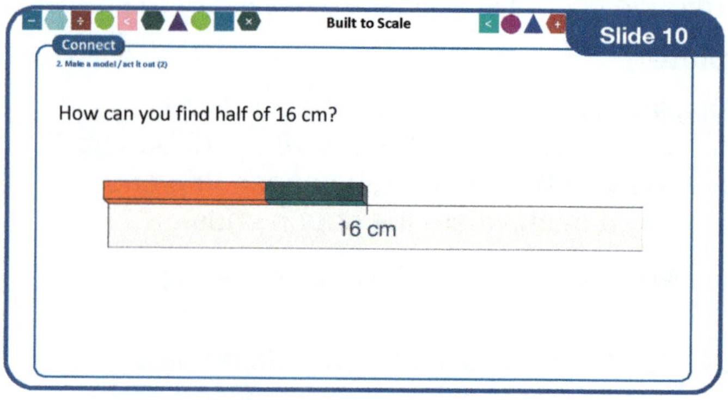

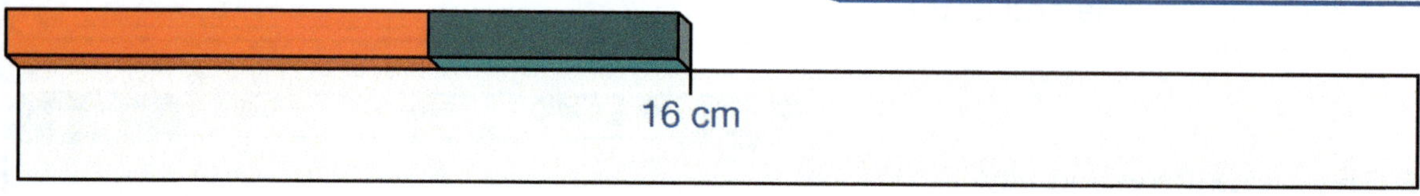

Look out for children who are not accurately lining up the end of the first number rod with the end of the paper strip.

- *How can you find half of 16 cm?*
- Two brown rods are the same length as the orange and dark green rod.

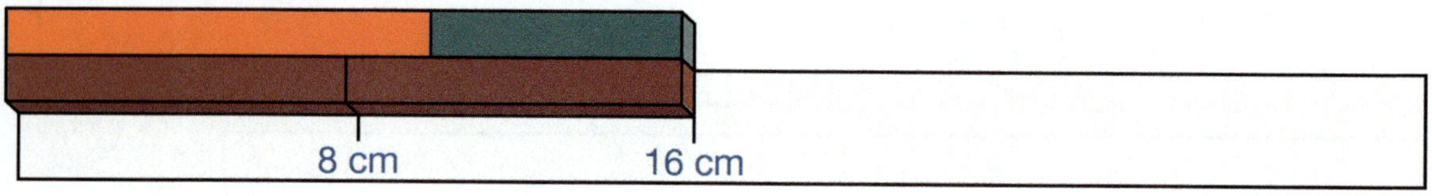

This process can be repeated to find 4 cm, 2 cm and 1 cm.

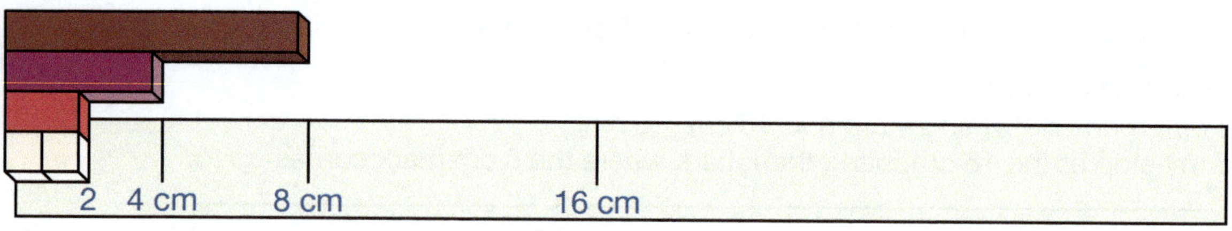

Then by combining rods, children will be able to mark other lengths on their ruler, including lengths greater than 16 cm.

Refer to the **Collaborate** task of the **Pupil Resource**.

Slide 11

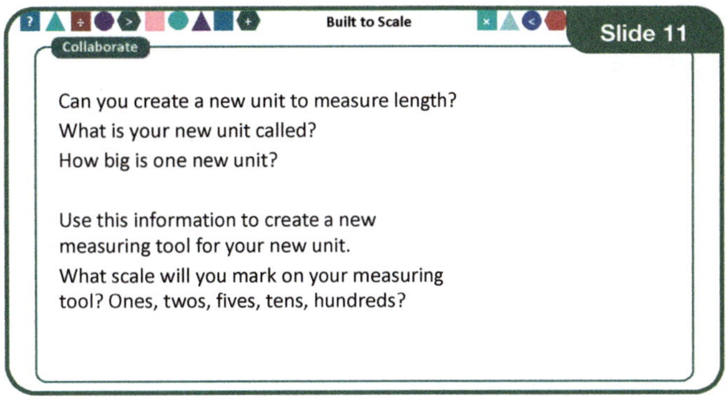

Display slide 11.

This task provides an opportunity for the children to work independently, in pairs or in small groups to create scales of their own.

Once the children have created their own rulers, challenge them to create their own measurement scales and an accurate tool to measure it with.

- *Can you create a new unit to measure length?*
- *What is your new unit called?*
- *How big is one new unit?*
- *Using this information, can you create a new measuring tool for your new unit?*
- *What scale will you mark on your measuring tool?*
- *How will you decide on your scale?*

Refer to the **Communicate** task of the **Pupil Resource**.

Slide 12

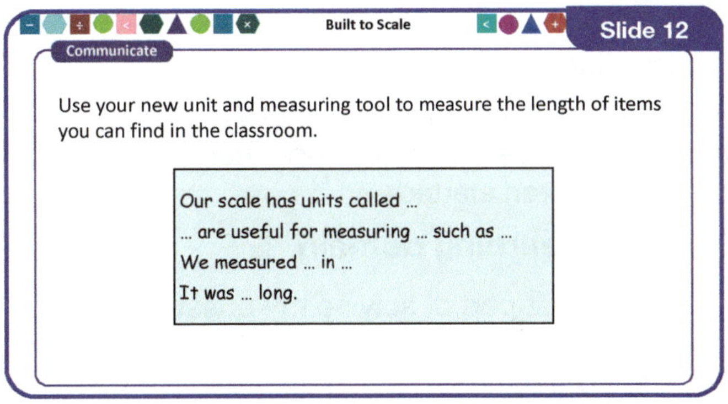

Display slide 12.

This task provides an opportunity for the children to compare the different units of measurement created: what is the same, and what is different? They can also measure items using each other's measurement units. They could consider what is appropriate to be measured in that unit. For example, if it is a small unit of measurement then it may be more useful for measuring smaller items rather than the length of the classroom.

- Our scale has units called …
- … are useful for measuring … such as …
- We measured the … in …
- It was … long.

For example:

Our scale has units called [wiggles].

[Wiggles] are useful for measuring [smaller items] such as [pencils].

We measured the [pencil case] in [wiggles]. It was [15 wiggles] long.

Create

Refer to the **Create** task of the **Pupil Resource**.

Slide 13

Display slide 13.

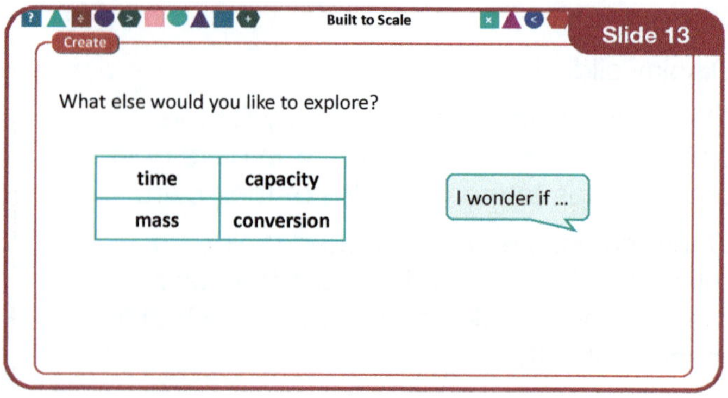

- *Were there any questions asked at the start of the task that you now want to think about?*
- *Is there anything new that you are wondering now?*

Children can use the word bank to prompt further questions of their own.

time	capacity
mass	conversion

- I wonder if I could make a new scale for capacity / mass / time.
- I wonder if I could create two scales that are conversions of each other (like centimetres and metres).

Consider

Refer to the **Consider** task of the **Pupil Resource**.

Slide 14

Display slide 14.

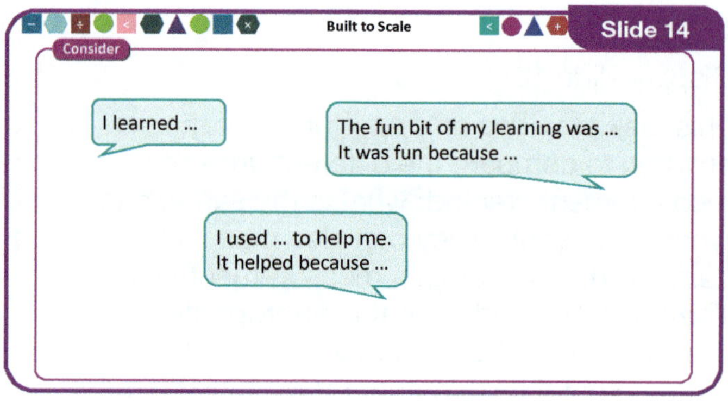

Encourage children to think about the strategies that they have used. The prompts below may help get children started.

Cognitive learning domain

- The trickiest part of solving my question was …
- I learned …
- The best strategy was …

Affective domain

- The fun bit of my learning was … It was fun because …
- I said 'aha' when … I said it because …
- I was surprised about … because …

Metacognitive domain

- I got stuck when … I unstuck myself by …
- I used … to help me. It was helpful because …
- Next time, I will …

Answers

The shortest ribbon is 4 cm long.

The longest ribbon is 18 cm long.

Peg Board Creations

Year 1

Introduction

Peg Board Creations provides children with an opportunity to explore properties of shapes. They will start by working with triangles but as they progress through the task they will begin to explore different shapes.

This task could be explored using a variety of problem-solving heuristics, ranging from making a model, to drawing a diagram or acting it out. The choices children make will determine which learning behaviours they will need to bring to the task.

The **Communicate** task requires children to apply understanding of positional language in order to accurately describe their models or diagrams to a partner. When explaining, they need to be precise. When listening, they will need to be able to translate instructions onto their own diagram with no visual prompt. If discrepancies are noted in the resulting drawings or models, children will have the opportunity to 're-think' their instructions to see how the error may have arisen.

This task promotes greater-depth learning in a range of ways, supporting children to identify and articulate underlying mathematical structures. For example, in the **Curious** stage, greater depth is promoted by supporting children to offer more than one good mathematical question.

This **Teacher Guide** is supported by photocopiable **Pupil Resources** and **Teacher's Presentation** PowerPoint slides for whole-class or group display.

National Curriculum

Peg Board Creations covers these Year 1 National Curriculum objectives.

Geometry – properties of shapes

- Recognise and name common 2D and 3D shapes, including:
  - 2D shapes [for example, rectangles (including squares), circles and triangles].
- *Pupils recognise these shapes in different orientations and sizes, and know that rectangles, triangles, cuboids and pyramids are not always similar to each other.* [non-statutory guidance]

Geometry – position and direction

- *Pupils use the language of position, direction and motion, including: left and right, top, middle and bottom, on top of, in front of, above, between, around, near, close and far, up and down, forwards and backwards, inside and outside.* [non-statutory guidance]

The 6C cycle

The grid on the right shows the relative weighting given to each of the 6 Cs in **Peg Board Creations**.

Curious	★★★
Connect	★
Collaborate	★
Communicate	★★
Create	★★★
Consider	★★

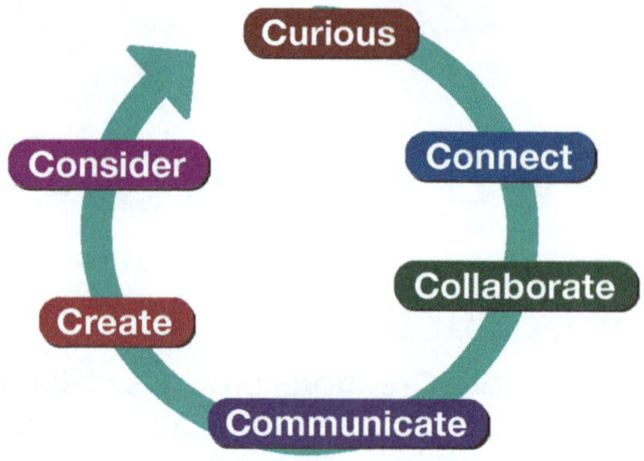

Problem-solving heuristics

The grid shows which problem-solving strategies children are likely to use when investigating the Connect tasks in **Peg Board Creations**. These strategies are further exemplified in the Connect stage below (pages 134–136).

Draw a diagram	✓
Make a table / spot a pattern	
Make a model / act it out	✓
Try then improve	✓
Create a systematic list	
Solve a simpler but related problem	
Work backwards	

Key mathematical ideas

This is an investigative task where children explore the three- and four-sided shapes they can create on a geoboard with a 3 by 3 array of pegs. The link between the number of vertices and the number of sides could arise in the 'Make a model / act it out' Connect task. It is important that children do not always create shapes with the same orientation, so encourage them to look from all angles.

In the Communicate task, children will need to use and listen to positional language (including combinations, such as top middle, bottom right), in order to both give and follow instructions.

Some children may choose to go on to explore different-sized grids.

Due to the explorative nature of the task, no answers are provided, although examples are given throughout.

Teaching guidance

Teaching guidance is given below.

The **Teacher's Presentation** PowerPoint file has slides specifically mentioned in this teaching guidance. Some of the slides include notes (within the PowerPoint file) providing 'stumble support' suggestions, that may help you to support children in their problem-solving thinking and enable access for children who are less confident.

Awe and wonder

Slide 1

Display slide 1.

Allow children to share what they can see and are noticing.

- I can see 9 dots.
- I can see a shape with 3 sides.
- I have noticed that the triangle has a dot in the middle.

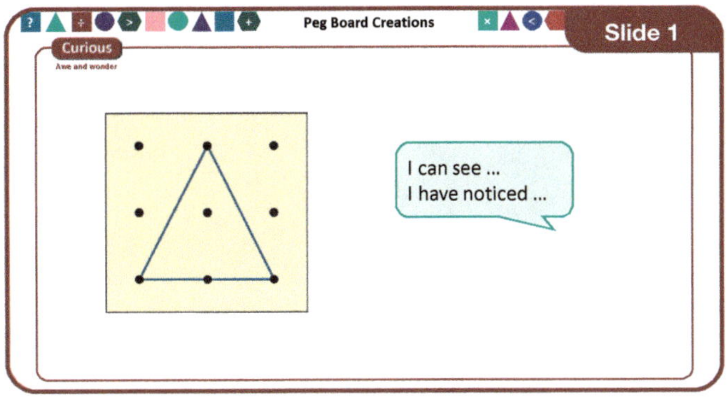

Develop and deepen

Slide 2

Display slide 2.

Encourage children to use the words in the word bank to develop and deepen their thoughts and questions.

join	**larger**
dots	**more**
triangle	**less**
sides	**smaller**

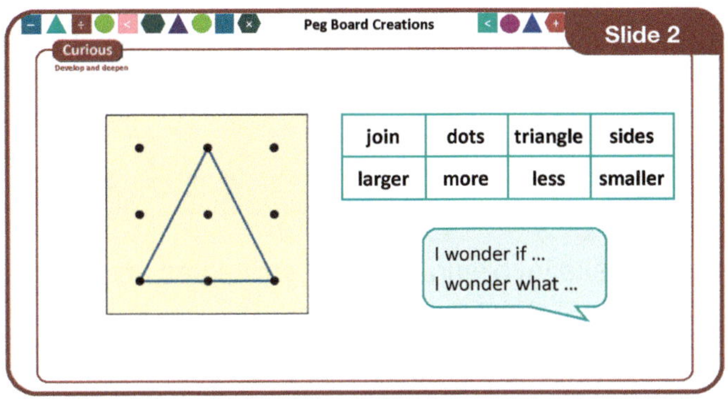

Possible lines of enquiry linked to the word bank:

- I wonder if I could make a different triangle.
- I wonder if I could make a smaller / larger triangle.
- I wonder if I could make triangles using all of the dots.
- I wonder if I could make more triangles inside the big triangle.
- I wonder if I could make other shapes by joining the dots.

Possible lines of enquiry that children may raise now, but should be held back until the **Create** task:

- I wonder if I could make more squares or more triangles by joining the dots.
- I wonder what shapes I could make if I used a grid with a different number of dots.

Consider whether / how to collate and display all the suggested questions, and whether / how to steer children towards selecting an appropriately challenging question.

Pupil Resources

Distribute the **Pupil Resources** for **Peg Board Creations** to the children to help support them in their exploration of the different tasks.

To provide additional flexibility in the classroom, each of the **Pupil Resources** is also provided in the **Teacher's Presentation** PowerPoint file (see slides 3–6), so you can display these at any time if you wish.

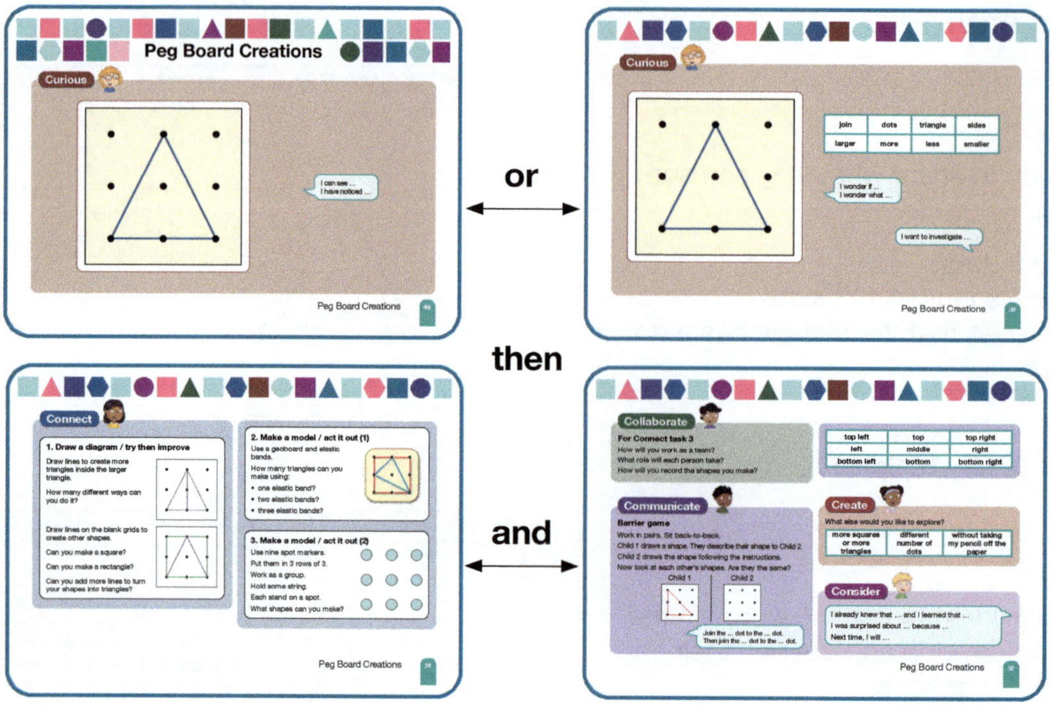

Three tasks involving different strategies are suggested for children to make connections.

1. Draw a diagram / try then improve

Additional resources required: rulers, coloured pencils

Slides 7 and 8

Print slides 7 and 8.

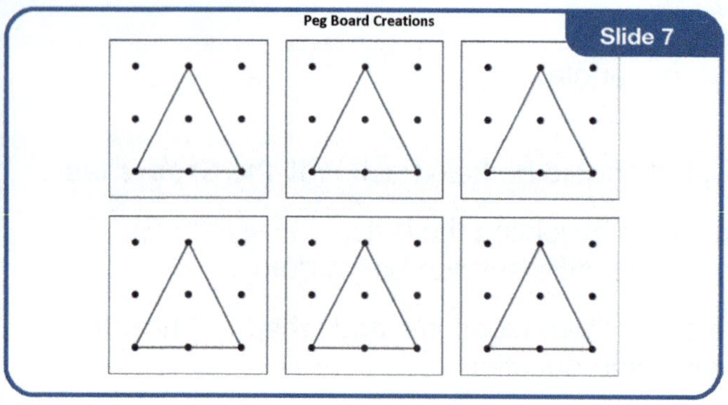

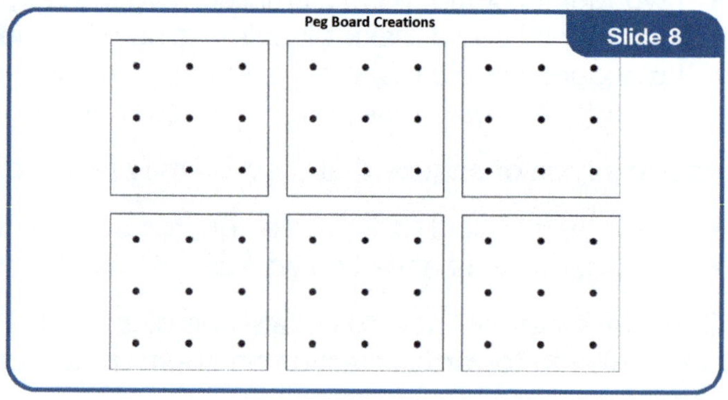

Slide 9

Display slide 9.

Distribute print-outs of slide 7 to the children. Children can draw lines on the grids to explore different ways to connect dots to create triangles within the larger triangle.

You could also cut the sheet into individual grids (or strips of three), so children can focus on one possibility (or one set of three linked possibilities) at a time.

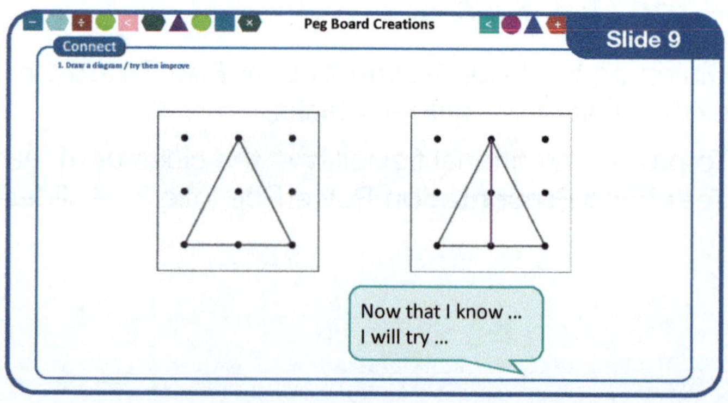

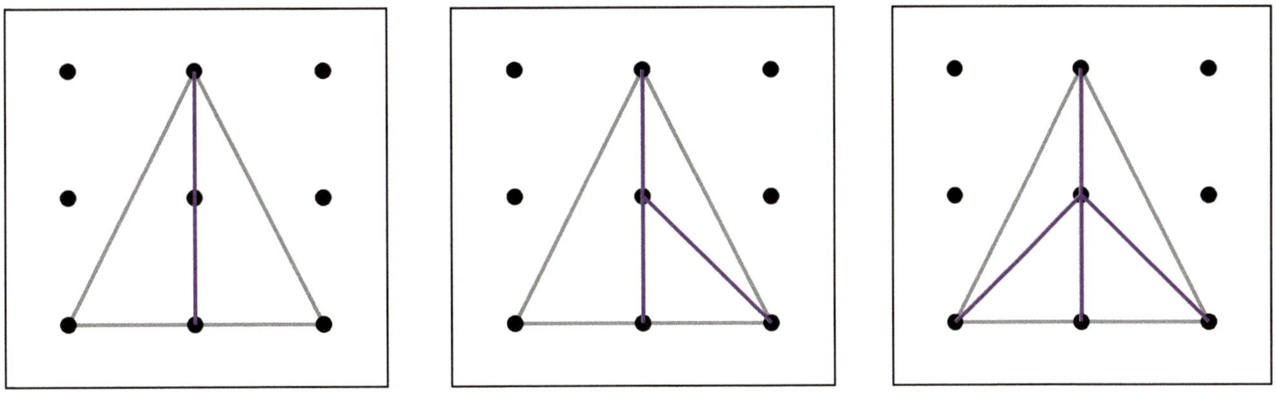

⚠ Look out for children who do not recognise triangles when they are presented in unusual orientations, or when they are less common types of triangle.

Slide 10

Display slide 10.

Distribute print-outs of slide 8 to the children. Children can draw lines on the grids to make other shapes.

- *Can you make a square or a rectangle?*
- *How many different squares and rectangles can you make?*
- *Now can you add more lines to turn your square / rectangle into triangles?*

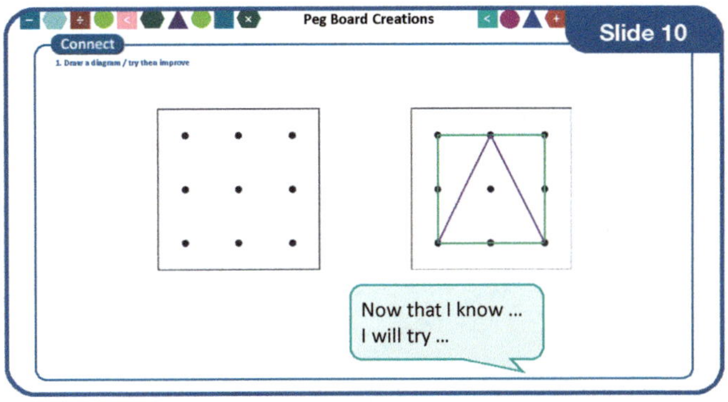

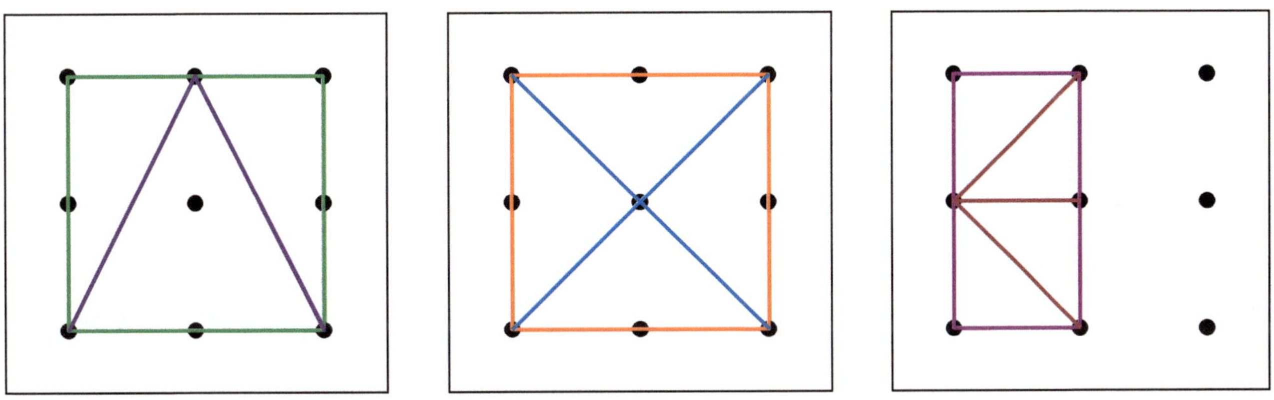

2. *Make a model / act it out (1)*

Additional resources required: geoboards and elastic bands (or a digital geoboard)

Slide 11

Display slide 11.

Children could use a geoboard and elastic bands to explore the different triangles that can be made.

- *How many triangles can you make using just one elastic band?*
- *How many triangles can you make using two or three elastic bands?*

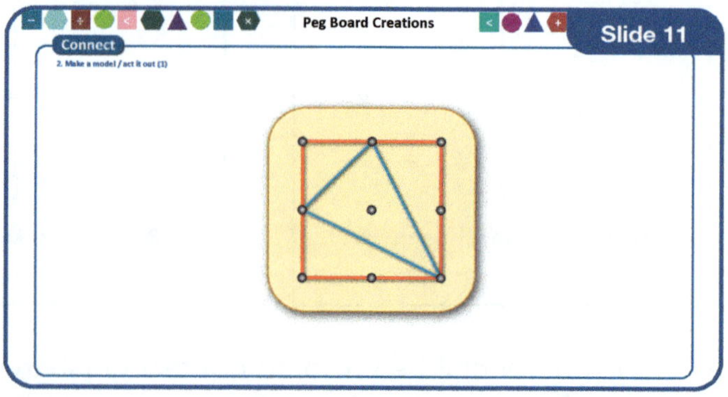

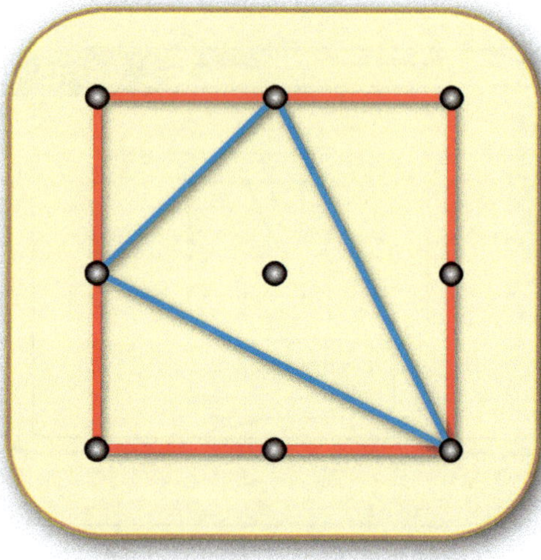

Example of four triangles created using two elastic bands on a geoboard

3. *Make a model / act it out (2)*

Additional resources required: spot markers, string

Slide 12

Display slide 12.

Use nine floor spot markers to mark out a grid on the floor (ideally you will have one grid per group of children, so a large outdoor space would be best). Ask the children to work in groups, and provide them with string or ropes to form the sides of the shapes.

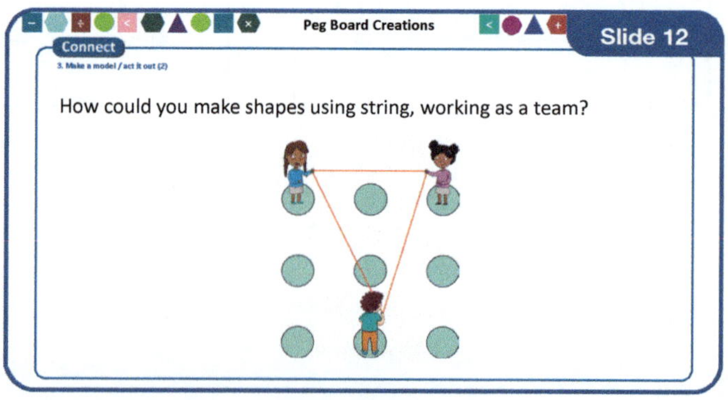

Refer to the **Collaborate** task of the **Pupil Resource**.

Connect task 3 gives children the opportunity to work in groups of three or four.

Some things for the groups to think about:

- *How many children does it take to make a triangle?*
- *In your group, can you make a four-sided shape?*
- *Has everyone got the same role?*
- *How will you record the shapes you create? Who will record them?*

Refer to the **Communicate** task of the **Pupil Resource**.

Additional resources required: geoboards and elastic bands (optional)

Slide 13

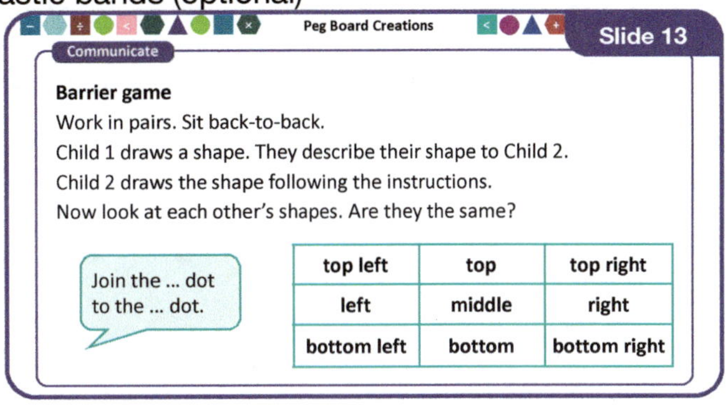

Display slide 13.

Give children more print-outs of slide 8 (the blank grids). Alternatively, they could use geoboards and elastic bands.

This task provides an opportunity for the children to learn and practise positional language, as well as giving and listening to instructions.

Children play the barrier game in pairs. They sit back-to-back, so they cannot see each other's drawing or model. They will take it in turns to draw a shape on the grid / model a shape on the geoboard and describe it to the other child, who then has to recreate it.

Some children might need more support with phrases like 'top left' and 'bottom right' as these are a combination of positional language.

Once the child listening to the instructions has finished, children can compare their drawings / models.

- *Are they the same? Are there any differences?*
- *If there are differences, which instruction led to this?*
- *How could your partner have described the shape more clearly?*

⚠ Look out for children who do not know their right from their left. There could be the need to create a mnemonic or a strategy to help children remember.

Create

Refer to the **Create** task of the **Pupil Resource**.

Slide 14

Display slide 14.

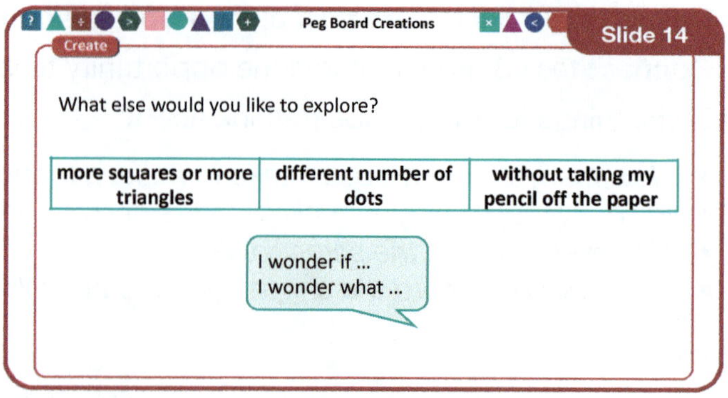

- *Were there any questions asked at the start of the task that you now want to think about?*
- *Is there anything new that you are wondering now?*

Children can use the word bank to prompt further questions of their own.

more squares or more triangles	**different number of dots**	**without taking my pencil off the paper**

- I wonder if I could make more squares or more triangles by joining the dots.
- I wonder what shapes I could make if I used a grid with a different number of dots.
- I wonder if I could join all of the dots to make triangles without taking my pencil off the paper.

For example, children might want to explore a 4 by 4 grid, a 3 by 4 grid or a 2 by 5 grid.

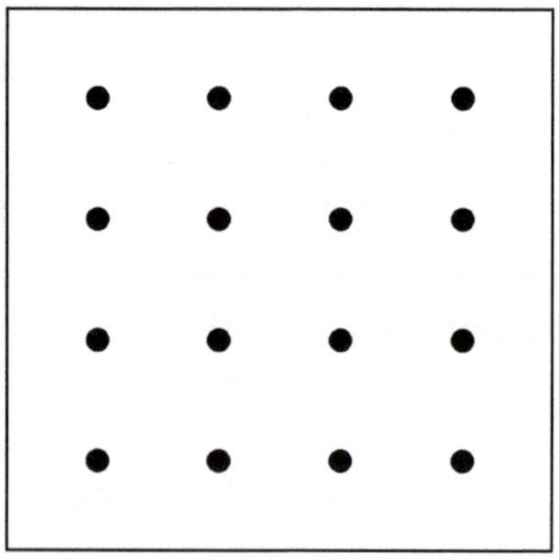

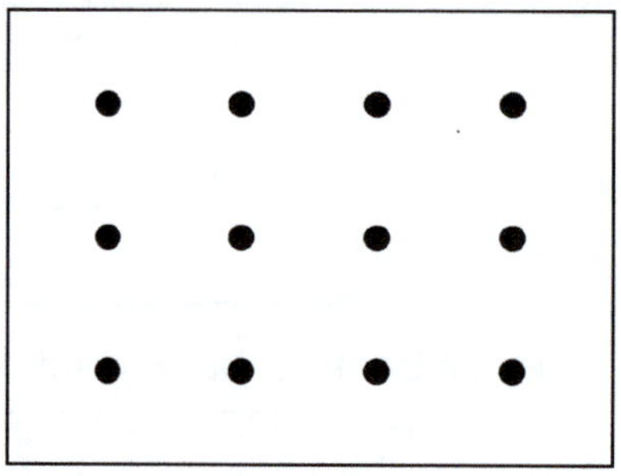

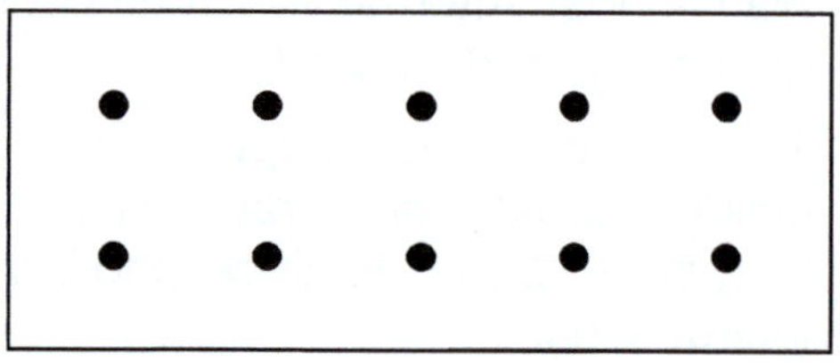

Refer to the Consider task of the **Pupil Resource**.

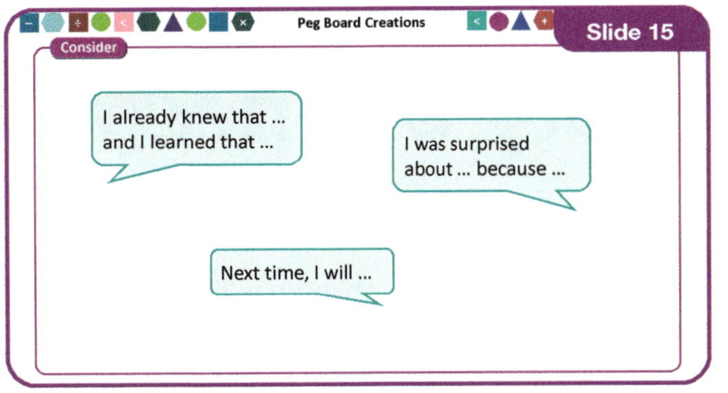

Slide 15

Display slide 15.

Encourage children to think about the strategies that they have used. The prompts below may help get children started.

Cognitive learning domain

- The trickiest part of solving my question was …
- I learned …
- I already knew that … and I learned that …

Affective domain

- The fun bit of my learning was … It was fun because …
- I said 'aha' when … I said it because …
- I was surprised about … because …

Metacognitive domain

- I got stuck when … I unstuck myself by …
- I used … to help me. It was helpful because …
- Next time, I will …

Story Shapes

Year 2

Introduction

Story Shapes provides children with the opportunity to apply and extend their knowledge of shapes, including their properties. Children will examine pictures created using tangram pieces and are encouraged to rehearse specific vocabulary linked to shapes and recognise common shapes in different orientations. Through exploration and manipulation of the tangram pieces, children can experiment with their own arrangements to create different pictures. These can be made more realistic by adding drawings and details to their creations.

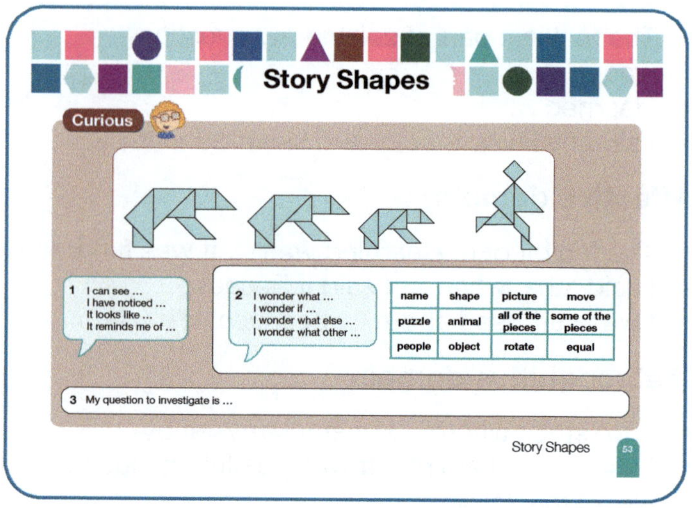

Children will be encouraged to work collaboratively, comparing and contrasting their ideas with others, and seeking information and clarity by asking respectful questions. During the **Communicate** task, children can continue to work together in order to recreate or create their own story using the pictures they have made.

This task promotes greater-depth learning in a range of ways, supporting children to identify and articulate underlying mathematical structures. For example, in the **Collaborate** stage, greater depth is promoted by supporting children to be polite and encouraging to group members, and to offer their own ideas and accept it if they are rejected by the group.

This **Teacher Guide** is supported by photocopiable **Pupil Resources** and **Teacher's Presentation** PowerPoint slides for whole-class or group display.

National Curriculum

Story Shapes covers these Year 2 National Curriculum objectives.

Geometry – properties of shapes

- Identify and describe the properties of 2D shapes, including the number of sides and line symmetry in a vertical line.
- Compare and sort common 2D and 3D shapes and everyday objects.
- *Pupils identify, compare and sort shapes on the basis of their properties and use vocabulary precisely, such as sides, edges, vertices and faces.* [non-statutory guidance]

The 6C cycle

The grid on the right shows the relative weighting given to each of the 6 Cs in **Story Shapes**.

Curious	★★★
Connect	★★
Collaborate	★★★
Communicate	★★
Create	★
Consider	★★

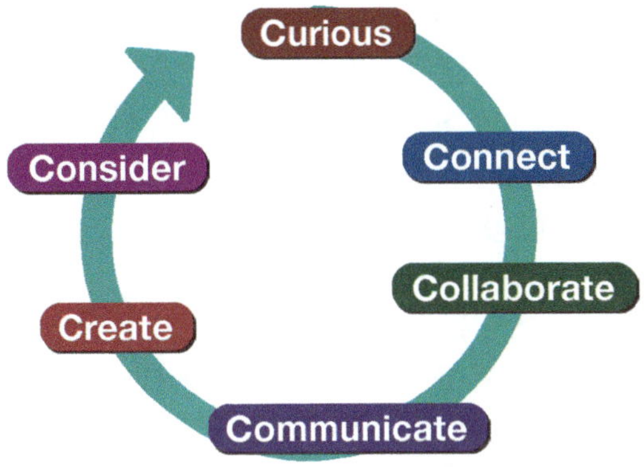

Problem-solving heuristics

The grid shows which problem-solving strategies children are likely to use when investigating the Connect tasks in **Story Shapes**. These strategies are further exemplified in the Connect stage below (pages 144–147).

Draw a diagram	✓
Make a table / spot a pattern	
Make a model / act it out	✓
Try then improve	
Create a systematic list	
Solve a simpler but related problem	
Work backwards	✓

Key mathematical ideas

A tangram is a Chinese puzzle which is split into 7 shapes. When placed together, the pieces make a square. Tangrams are made up of 5 triangles (2 large, 1 medium and 2 small), a square and a parallelogram.

There is no expectation for children in Key Stage 1 to use the name or describe a parallelogram. However, they should be encouraged to describe what they notice about the shape and make comparisons with shapes that they know by name. For example, they may make connections to a rectangle, noticing how the shapes are similar and how they differ. In Key Stage 2, children will explore that parallelograms have: two pairs of equal, straight sides; opposite sides that are parallel; opposite angles that are equal; interior angles that total 360°.

Children may recognise that some shapes within a tangram are the same, for example, there are two large, identical triangles. The mathematical term for this is 'congruent'. Congruent shapes have identical properties including length of sides and angles, but may be displayed in different orientations. There is no expectation for children in Key Stage 1 to be introduced to this term. Simply encourage them to see simple shapes in different orientations and recognise that they are the same. At this stage in their learning, they should understand that a shape can change orientation when moved, but remain the same size and shape.

Children should be encouraged to use the following vocabulary, as well as accurate shape names, when discussing and describing the shapes that they have used to represent different objects.

To describe shapes: shape, flat, side, vertex (plural vertices), line of symmetry.

To describe lines: line, curved, straight, round.

To name shapes: square, triangle, rectangle (children may draw similarities and differences between a rectangle and a parallelogram),

Vocabulary has been chosen to reflect working with the shapes within the tangram puzzle, so not all expected vocabulary for properties of shapes has been included above.

Slide 9 provides images of potential tangrams suitable for the story of *Goldilocks and the Three Bears*. Children could imitate the tangrams, first considering the positioning and orientation of shapes before creating their own.

Due to the explorative nature of the task, no answers are provided, although examples are given throughout.

Teaching guidance

If you are using the **Express Pathway**, please turn to pages 144–148.

Teaching guidance for the **Full 6C Pathway** is given below.

The **Teacher's Presentation** PowerPoint file has slides specifically mentioned in this teaching guidance. Some of the slides include notes (within the PowerPoint file) providing 'stumble support' suggestions, that may help you to support children in their problem-solving thinking and enable access for children who are less confident.

Awe and wonder

Reveal slides 1–3 one at a time, allowing time for discussion and observations as prompted below.

Slide 1

- I can see some shapes.
- I have noticed that the shapes make a picture.
- It looks like a bear.
- It reminds me of a jigsaw puzzle.
- I have noticed some triangles / a square in the picture.
- I can see a shape that I don't know the name of.

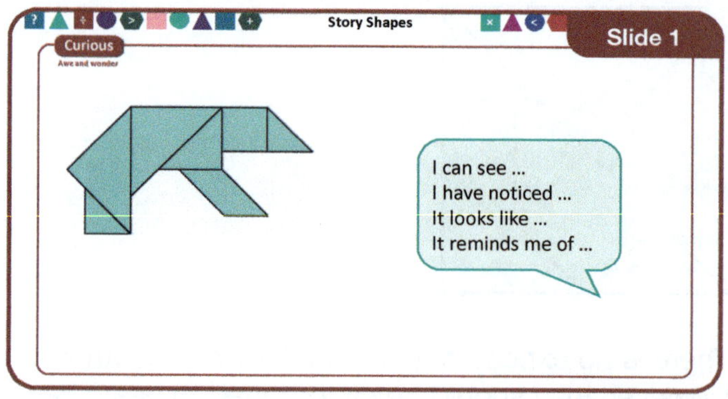

Slide 2

- I can see three bears now, but they are different sizes.
- I have noticed that all the bear pictures use the same shapes.

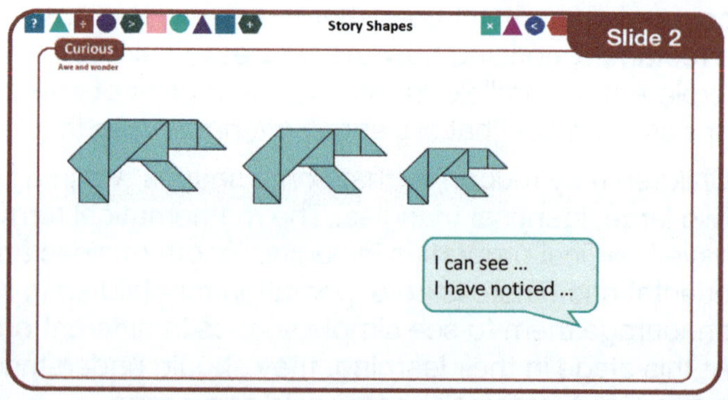

Slide 3

- I can see three bears and a person.
- It looks like a person running.
- It reminds me of *Goldilocks and the Three Bears*.
- I have noticed that the same shapes are used in all of the pictures, but they are different sizes, and the shapes in the last picture are turned around.

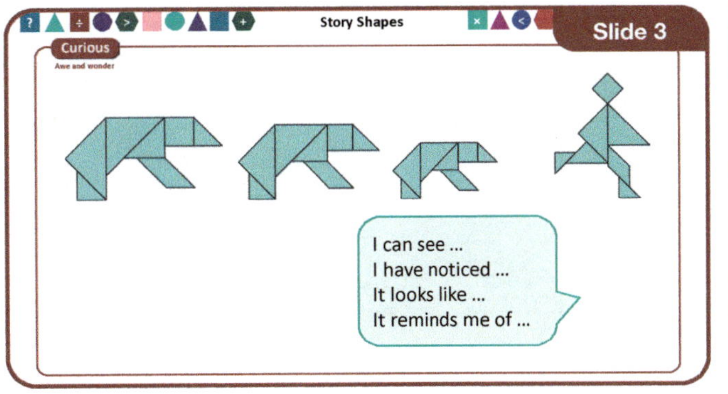

Develop and deepen

Encourage children to use the words in the word bank to develop and deepen their thoughts and questions.

name	**shape**	**picture**	**move**
puzzle	**animal**	**all of the pieces**	**some of the pieces**
people	**object**	**rotate**	**equal**

Possible lines of enquiry linked to the word bank:

- I wonder what the name of this shape is.
- I wonder if I could make a different animal using some of the pieces.
- I wonder what other pictures I could make by moving / rotating the shapes.
- I wonder if I could make any other pictures from the story of *Goldilocks and the Three Bears*.
- I wonder what else I could make using some of the puzzle pieces.
- I wonder what other animals I could make using all of the puzzle pieces.

Possible lines of enquiry that children may raise now, but should be held back until the **Create** task:

- I wonder if I could make a similar puzzle with different-shaped pieces.
- I wonder if I could make pictures using seven different-sized triangles.

Consider whether / how to collate and display all the suggested questions, and whether / how to steer children towards selecting an appropriately challenging question.

Once children have come up with their lines of enquiry, explain what a tangram is.

- *A tangram is a Chinese puzzle which is split into 7 shapes. When placed together, the pieces make a square.*

Slide 8 could be used to show how the pieces fit together to make a square.

Pupil Resources for Full and Express Pathways

Distribute the **Pupil Resources** for **Story Shapes** to the children to help support them in their exploration of the different tasks. Use either of the **Curious Pupil Resources** depending on whether you are following the **Full Pathway** or the **Express Pathway**. The **Pupil Resources** for the remaining 5Cs are common to both pathways.

To provide additional flexibility in the classroom, each of the **Pupil Resources** is also provided in the **Teacher's Presentation** PowerPoint file (see slides 4–7), so you can display these at any time if you wish.

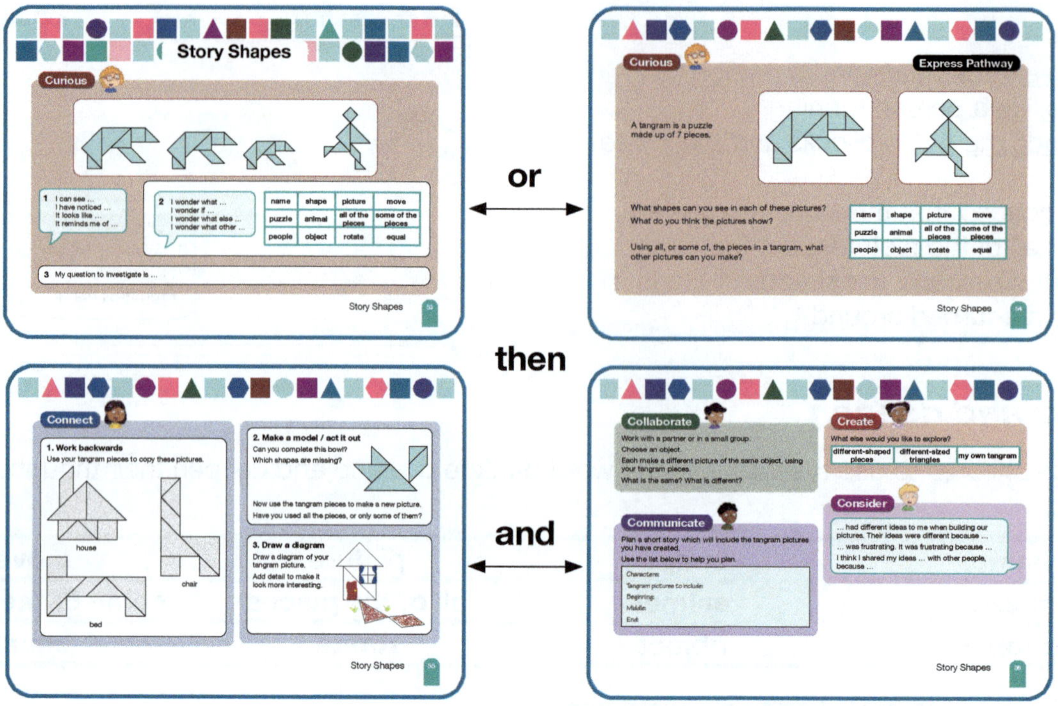

Express Pathway

If you are following the **Express Pathway**, spend a few minutes discussing the **Curious Pupil Resource** with the children. (This is slide 5, so you could display it if you wish.) Use some of the prompts on the previous page to help.

Three tasks involving different strategies are suggested for children to make connections.

1. Work backwards

Additional resources required: scissors

Slide 8

Print slide 8 and distribute to children. Ask them to carefully cut out the shapes.

Can they name the individual shapes?

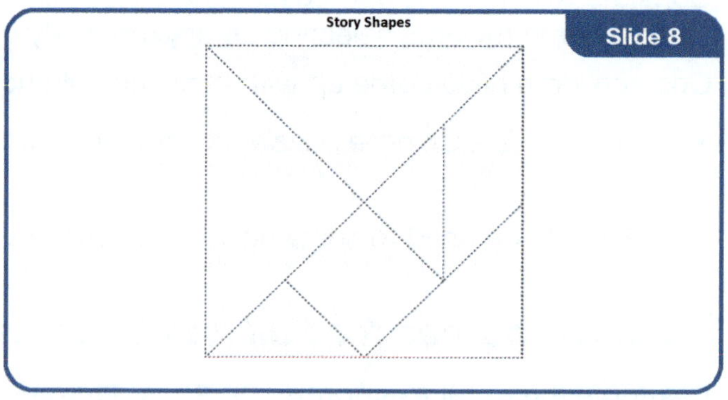

Slide 9

Print slide 9 and distribute to children.

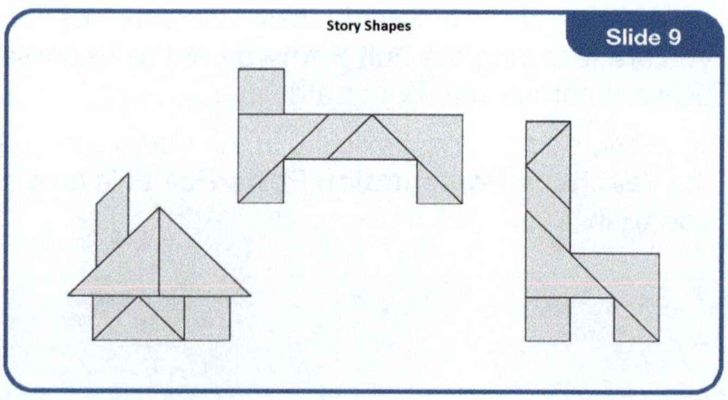

Slide 10

Display slide 10.

Ask children to work backwards to imitate the house picture using their cut-out tangram pieces.

Then ask them to do the same for the bed and chair pictures.

⚠ Watch out for any children who do not consider flipping the shapes over or rotating them, and are only using the shapes in their standard orientation.

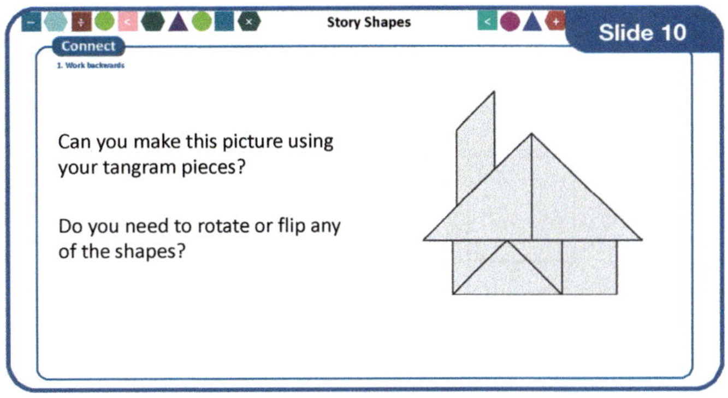

2. Make a model / act it out

Slide 11

If you would like to give children a greater selection of shapes to work with (same tangram, but in two different sizes), print slide 11 and distribute to children. (Also distribute fresh copies of slide 8 if needed.) Ask them to carefully cut out the shapes.

⚠ You may want to provide some pre-cut tangrams for children who are struggling with cutting.

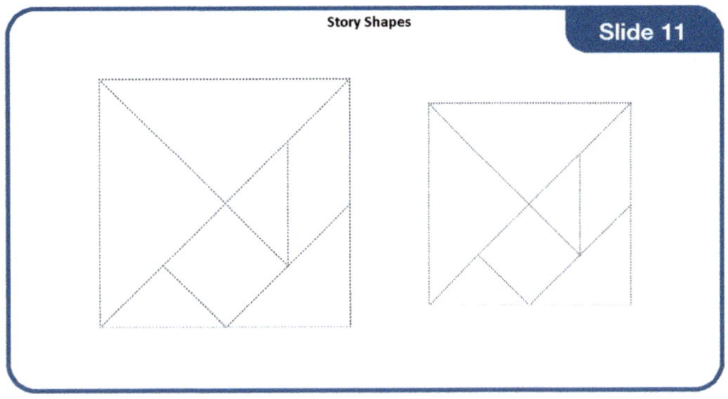

Challenge the children to make pictures using all or some of the shapes from the tangram. Their pictures could link to the story of *Goldilocks and the Three Bears*, or they could link to their own stories.

Children may start by making very simple pictures using only one or two pieces. For example:

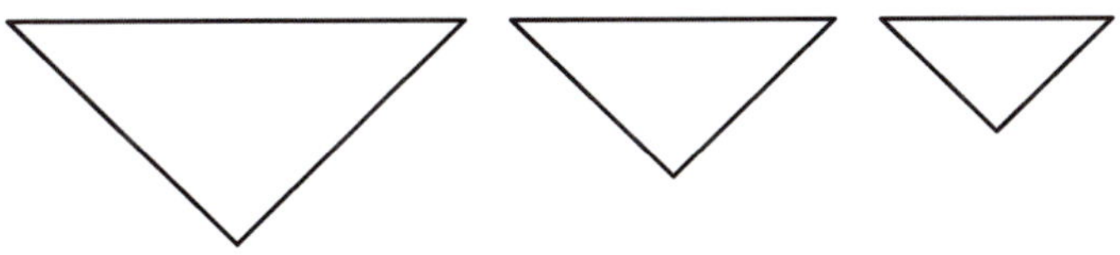

- These three triangles are porridge bowls. The largest is daddy bear's, the medium size is mummy bear's and the smallest is baby bear's.

Encourage children to try to make more complex pictures by using more shapes and by moving them around, rotating them and flipping them. Ask them to describe how they have used the shapes to create their pictures.

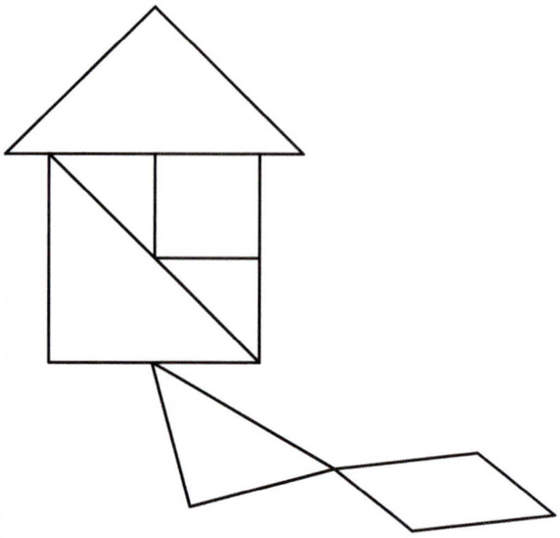

- I have made a new house for the bears. One of the big triangles is the roof. The square looks a bit like a window. Then I have made a path with two of the other shapes.

Slide 12

When children have had time to try out a few pictures, display slide 12.

- *I want to make a porridge bowl using all seven of the tangram pieces, but I can't quite finish it. Do you know which pieces are missing and where they go to complete the bowl?*

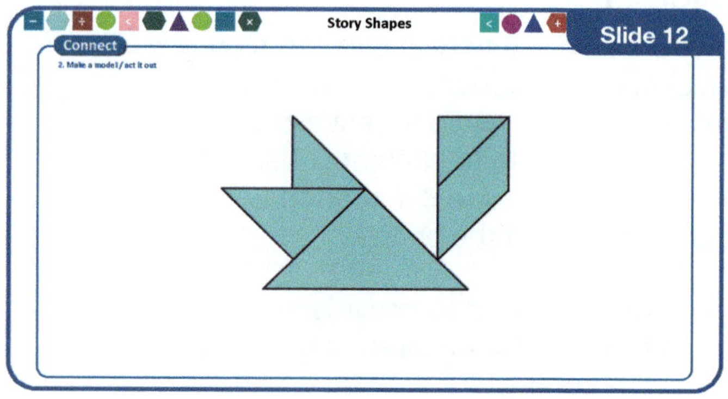

The square and one of the large triangles are missing. The completed bowl picture looks like this:

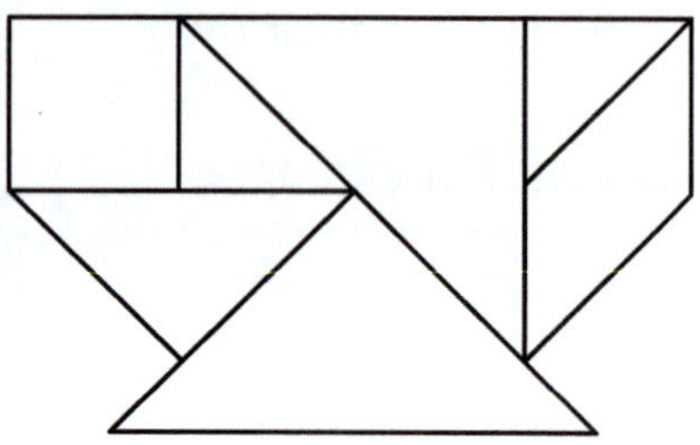

- *Are there other ways of making a bowl? Can you try your own version?*

3. Draw a diagram

Additional resources required: coloured pencils, glue (optional)

Slide 13

Display slide 13.

- *This is daddy bear's porridge bowl.*
- *How can we make this look even more like a bowl?*

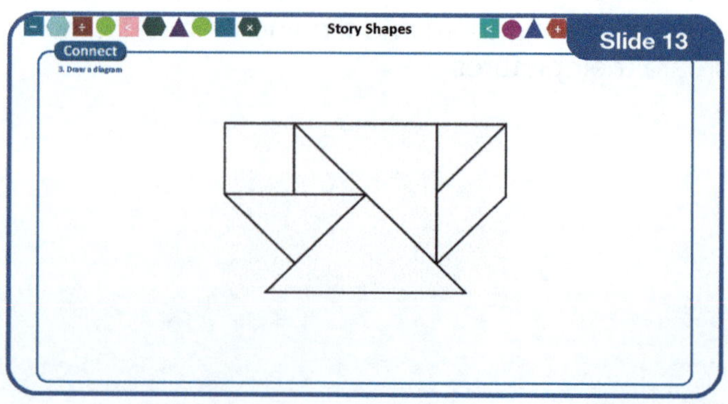

Model adding detail (for example, a pattern on the side of the bowl, and a spoon) by drawing on the picture.

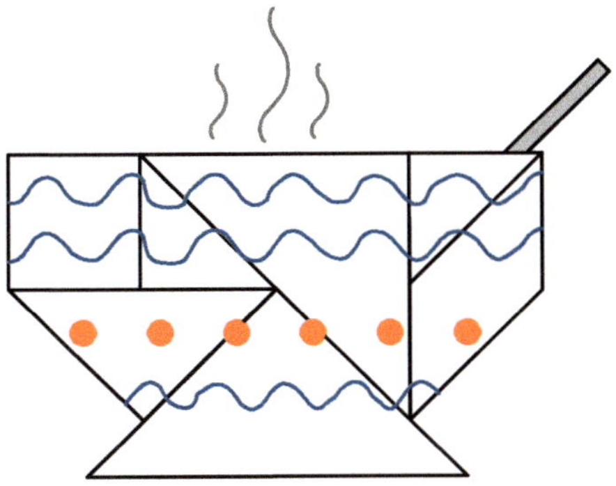

Once children have created a picture that they are happy with, ask them to make a diagram of what they have created. They could draw it freehand, draw around their tangram pieces, or glue their shapes onto a sheet of paper. Encourage them to add details to their diagram.

- I've turned the square into a window and added a front door. I've made the path look like stones and drawn some bits of grass.

Refer to the **Collaborate** task of the **Pupil Resource**.

Let the children work in pairs or small groups. Ask them to create a list of objects and then each try and find a different way to make each object using their tangram pieces. For example, they could all try making a dog in a different way.

- *How many different ways have you found to make the object?*
- *Have you used all the pieces or only some of them?*
- *What is the same?*
- *What is different?*

Communicate

Refer to the **Communicate** task of the **Pupil Resource**.

Slide 14

Display slide 14.

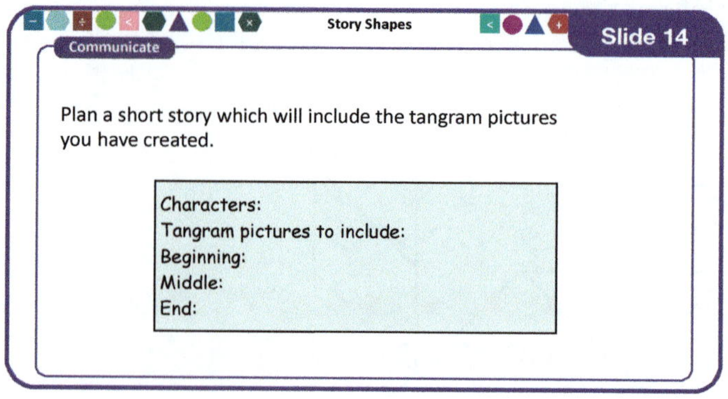

This task provides an opportunity for the children to work independently, in pairs or in small groups to write a short story including their tangram puzzles.

- Children could use characters and objects from *Goldilocks and the Three Bears*.
- Following their exploration of tangrams, children could write their own story based on the pictures they created.
- Digital images could be taken of their objects and these could be used to create a storyboard.

The children should use story language as appropriate and the word prompts on slide 14 to help them.

Create

Refer to the **Create** task of the **Pupil Resource**.

- *Were there any questions asked at the start of the task that you now want to think about?*
- *Is there anything new that you are wondering now?*

Children can use the word bank to prompt further questions of their own.

different-shaped pieces	**different-sized triangles**	**my own tangram**

- I wonder if I could make a similar puzzle with different-shaped pieces.
- I wonder if I could make pictures using seven different-sized triangles.
- I wonder if I could cut up a square into seven different shapes to create my own tangram.

Consider

Refer to the **Consider** task of the **Pupil Resource**.

Encourage children to think about the strategies that they have used. The prompts below may help get children started.

Cognitive learning domain

- The trickiest part of my investigation was … It was the trickiest because …
- I learned … I think this will be useful because …
- … had different ideas to me when building our pictures. Their ideas were different because …

Affective domain

- … was frustrating. It was frustrating because …
- I would change … about my investigation, because …
- I was surprised about … because …

Metacognitive domain

- I think I shared my ideas … with other people, because …
- Next time, I will …

Pattern People

Year 2

Introduction

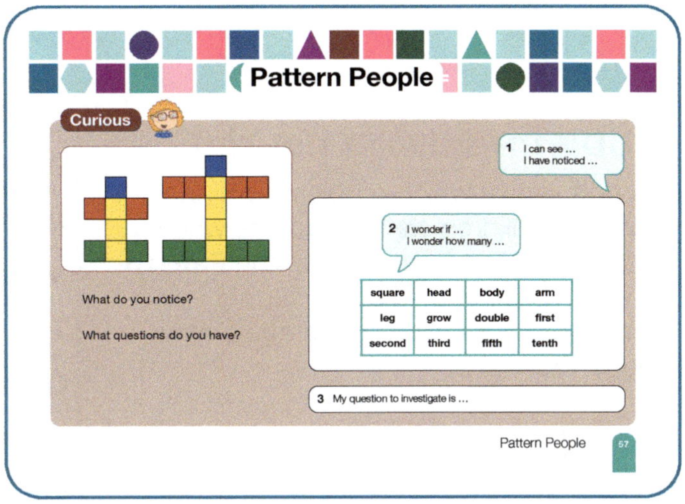

Pattern People provides children with an opportunity to explore a growing pattern, thinking about which elements remain constant and which change each time. The patterns are simple 'people' shapes, made up of different-coloured squares representing the head, body, arms and legs. Children will explore how the numbers of squares change as the people grow. A variety of problem-solving heuristics could be used to enable children to spot the patterns.

Children will need to work systematically, thinking about the best way to record their information either in tables or tally charts. The use of block diagrams or bar charts will help them to visualise the patterns and describe generalisations. They may begin to consider a relationship between the position of the person in the sequence and the number of squares they are built from.

This task promotes greater-depth learning in a range of ways, supporting children to identify and articulate underlying mathematical structures. For example, in the **Create** stage, greater depth is promoted by supporting children to form their own questions that are creative and inventive.

This **Teacher Guide** is supported by photocopiable **Pupil Resources** and **Teacher's Presentation** PowerPoint slides for whole-class or group display.

National Curriculum

Pattern People covers these Year 2 National Curriculum objectives.

Statistics

- Interpret and construct simple pictograms, tally charts, block diagrams and simple tables.
- Ask and answer simple questions by counting the number of objects in each category and sorting the categories by quantity.
- Ask and answer questions about totalling and comparing categorical data.

The 6C cycle

The grid on the right shows the relative weighting given to each of the 6 Cs in **Pattern People**.

Curious	★
Connect	★★
Collaborate	★★
Communicate	★★
Create	★★★
Consider	★★

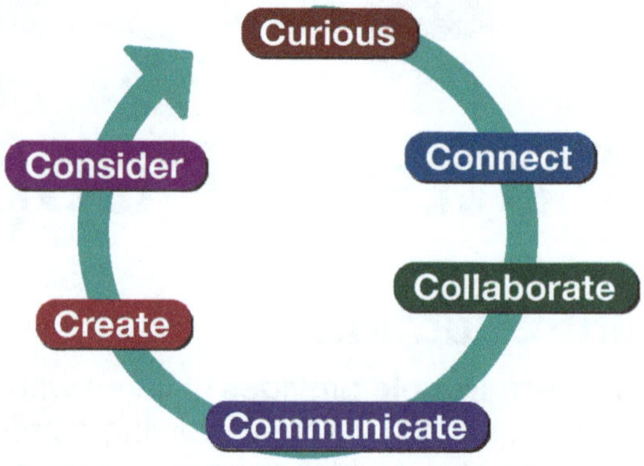

Problem-solving heuristics

The grid shows which problem-solving strategies children are likely to use when investigating the Connect tasks in **Pattern People**. These strategies are further exemplified in the Connect stage below (pages 153–157).

Draw a diagram	✓
Make a table / spot a pattern	✓
Make a model / act it out	✓
Try then improve	
Create a systematic list	
Solve a simpler but related problem	✓
Work backwards	

Key mathematical ideas

In this growing pattern, the head size remains constant as the person grows. The lengths of the body, arms and legs all increase. There is one additional 'body' square, two additional 'leg' squares (one for each side), and two additional 'arm' squares each time.

Children may notice other patterns as the people grow. For example:

- each time, 5 more squares are added (1 for the body, 2 for the legs and 2 for the arms);
- adding 2 to the 'person' number gives the body length (for example, person 1 has a body length of 3, and person 5 has a body length of 7);
- doubling the 'person' number gives the length of both arms (for example, person 1 has arms that are each 1 square long, and person 5 has arms that are each 5 squares long).

Through translating their data into graphs, the patterns become more obvious.

Children could use either block diagrams or bar charts. They are likely to need support in progressing from one to the other, as they get used to the differences. For example:

- block diagrams do not have spaces between the different categories; bar charts do;
- in block diagrams each individual unit in a category is visible; in bar charts, there is one solid unit per category;
- on the vertical axis of block diagrams the blocks are numbered; in bar charts, the scale is labelled on the division between the squares.

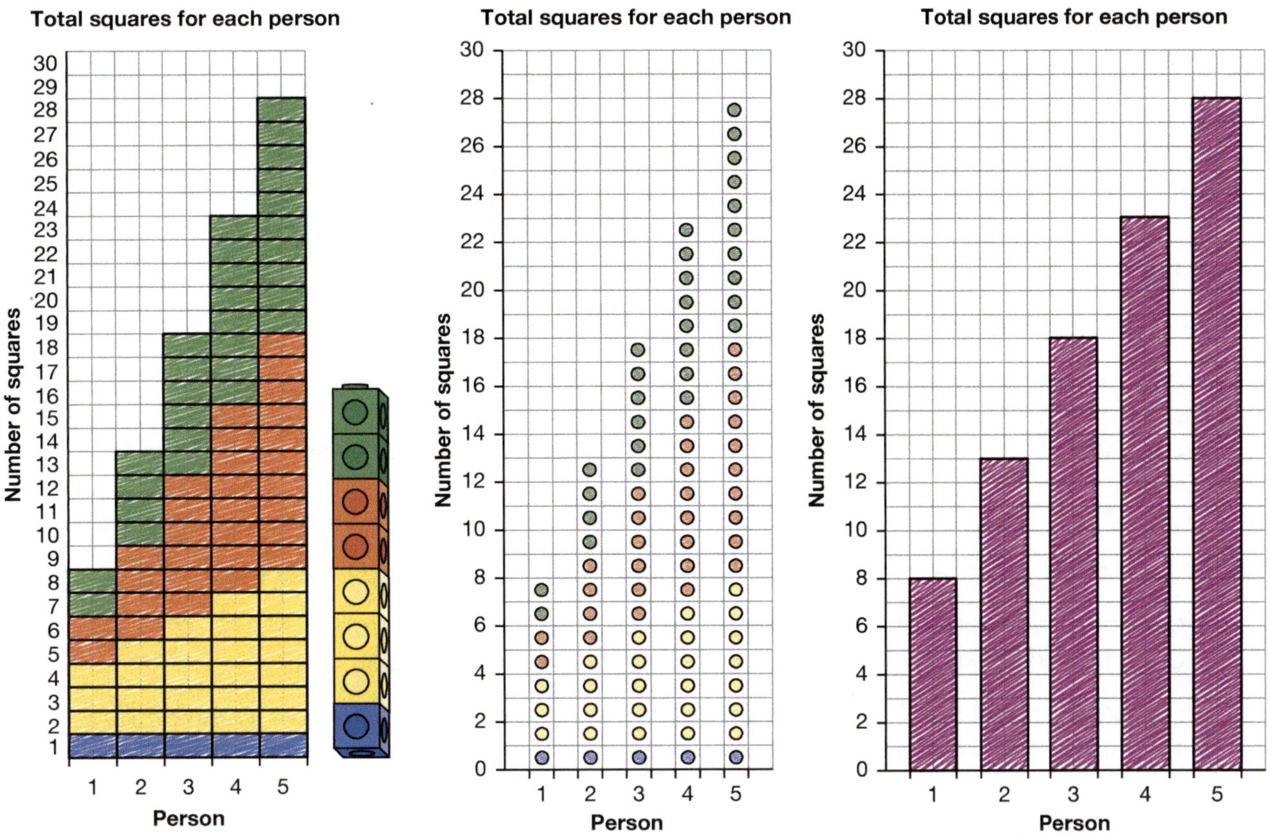

An example of progression from a block diagram (supported by concrete models) to a bar chart.

Answers are given in situ throughout the tasks because of the cumulative nature of the tasks.

Teaching guidance

If you are using the **Express Pathway**, please turn to pages 153–158.

Teaching guidance for the **Full 6C Pathway** is given below.

The **Teacher's Presentation** PowerPoint file has slides specifically mentioned in this teaching guidance. Some of the slides include notes (within the PowerPoint file) providing 'stumble support' suggestions, that may help you to support children in their problem-solving thinking and enable access for children who are less confident.

Awe and wonder

Slide 1

Display slide 1.

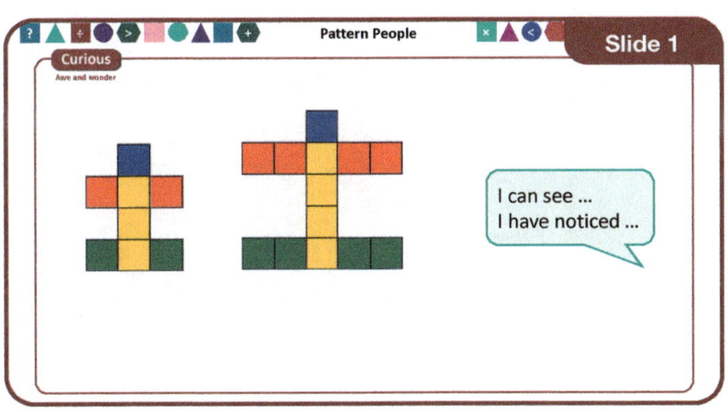

- I can see lots of squares.
- I have noticed that they look a bit like people.
- I can see that each person has a blue head, a yellow body, orange arms and green legs.
- I have noticed that one is smaller.
- I have noticed that the arms in the first picture are 1 square long, but in the second picture, they are 2 squares long.

Develop and deepen

Encourage children to use the words in the word bank to develop and deepen their thoughts and questions.

square	head	body	arm
leg	grow	double	first
second	third	fifth	tenth

Possible lines of enquiry linked to the word bank:

- I wonder how many squares long the third person's arms would be.
- I wonder how many squares the fifth person would be made from.
- I wonder if the tenth person would be made from double the number of squares as the fifth person.
- I wonder if there is a pattern in the number of squares each person is made of, as they grow.

Possible lines of enquiry that children may raise now, but should be held back until the **Create** task:

- I wonder how many squares the hundredth person would be made from.
- I wonder if I could make a growing pattern of animals made out of squares.

Consider whether / how to collate and display all the suggested questions, and whether / how to steer children towards selecting an appropriately challenging question.

Pupil Resources for Full and Express Pathways

Distribute the **Pupil Resources** for **Pattern People** to the children to help support them in their exploration of the different tasks. Use either of the **Curious Pupil Resources** depending on whether you are following the **Full Pathway** or the **Express Pathway**. The **Pupil Resources** for the remaining 5Cs are common to both pathways.

To provide additional flexibility in the classroom, each of the **Pupil Resources** is also provided in the **Teacher's Presentation** PowerPoint file (see slides 2–5), so you can display these at any time if you wish.

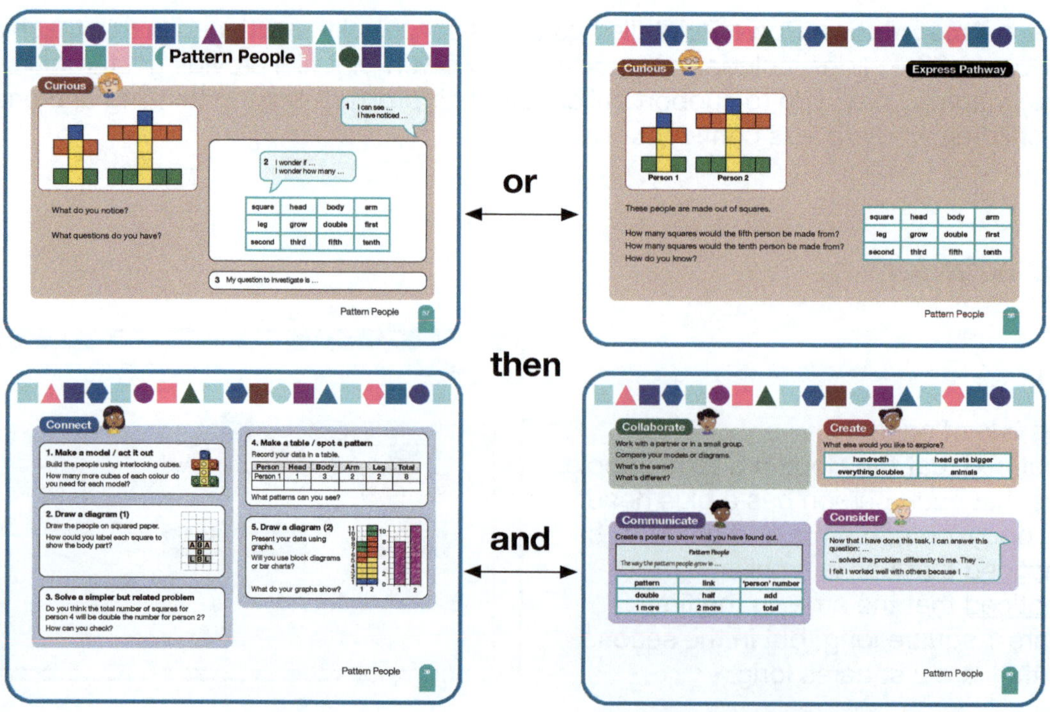

Express Pathway

If you are following the **Express Pathway**, spend a few minutes discussing the **Curious Pupil Resource** with the children. (This is slide 3, so you could display it if you wish.) Use some of the prompts on the previous page to help.

Five tasks involving different strategies are suggested for children to make connections.

1. Make a model / act it out

Additional resources required: interlocking cubes

Slide 6

Display slide 6.

Children could build models of the growing people in two ways:

1) they could start with person 1 and add on cubes to grow their original model each time;
2) they could build a new model each time for a direct comparison.

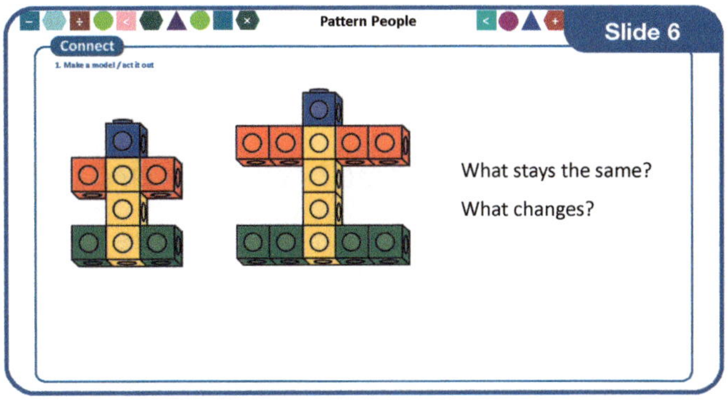

Models of the first three people, built using interlocking cubes

⚠ Ensure that children know the difference between cubes and squares. Do they recognise that they can use the square face of the cube to represent the squares that make up the people?

2. Draw a diagram (1)

Additional resources required: squared paper, coloured pencils (optional)

This task could be a follow-up to making models, or some children may prefer to move straight to drawing diagrams.

Slide 7

Display slide 7.

Encourage children to draw the people on squared paper.

- *What would be a quick way of labelling the squares that make up each body part?*
- I could write a letter H for head, B for body, A for arm and L for leg.

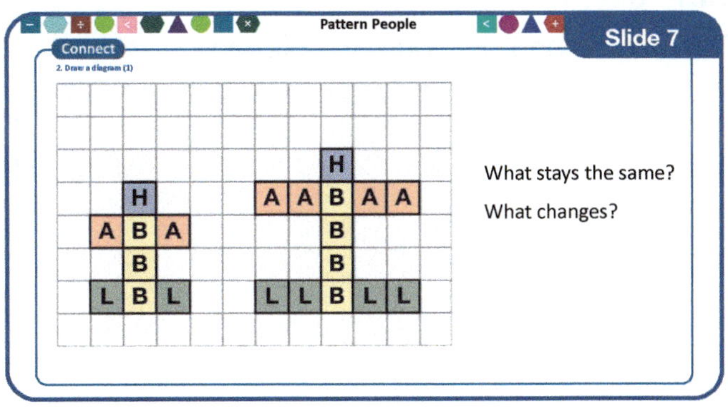

Some children may also wish to colour the squares so they can see at a glance which body part each square belongs to.

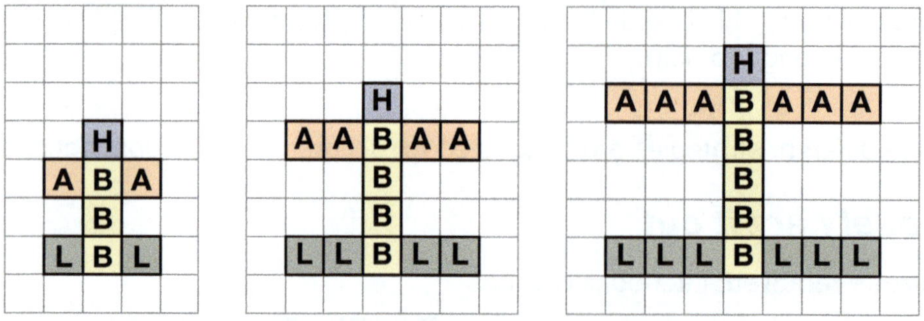

Diagrams of the first three people, drawn on squared paper

3. *Solve a simpler but related problem*

This task may be useful for children who are wondering whether person 10 would be made from double the number of squares as person 5.

- *What is double 2?* (4)
- *Do you think that means that person 4 will be made of double the number of squares as person 2?*
- *How could you check? How could you record what you find so you can prove it to someone else?*

Children could either make models, draw diagrams or list the numbers of squares in a table. For example:

Person	Head squares	Body squares	Arm squares	Leg squares	Total squares
Person 2	1	4	4	4	**13**
Person 4	1	6	8	8	**23**

- *Why is person 4 not made up of double the number of squares as person 2?*
- Because the head is always 1 square. And the body starts at 3 squares for person 1 and only goes up by 1 square each time.
- *Where can you see numbers that double?*
- The numbers of arm and leg squares for person 4 are double the numbers for person 2.
- *Now that you have done this, do you think the number of squares for person 10 will be double the number for person 5? Why / why not?*

4. *Make a table / spot a pattern*

Slide 8

Print slide 8 and distribute to children.

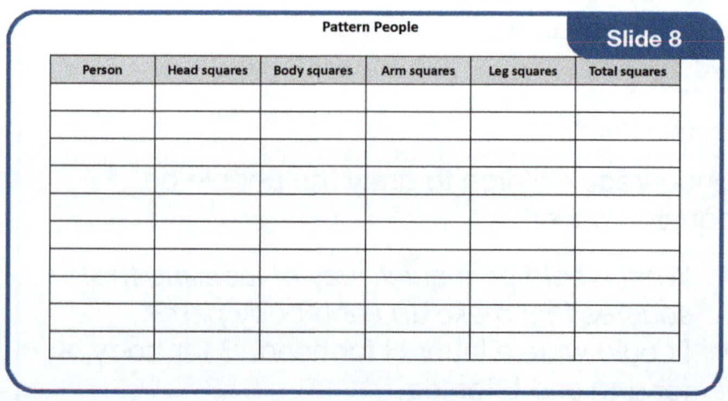

Slide 9

Display slide 9, which shows how to complete the first row for person 1.

Children could use the table format to record the information they discover through building models, drawing diagrams or visualising the growing people.

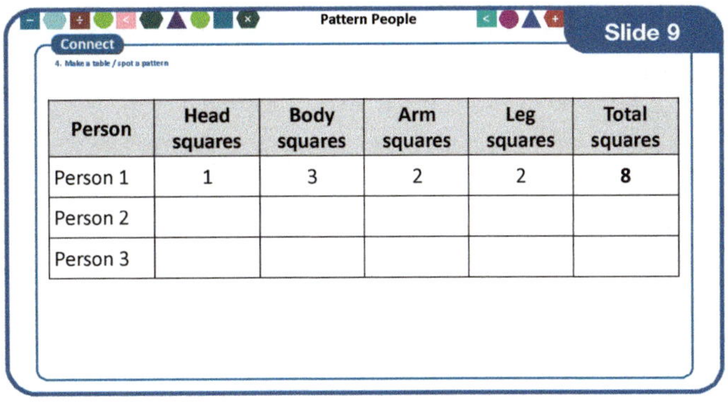

Person	Head squares	Body squares	Arm squares	Leg squares	Total squares
Person 1	1	3	2	2	**8**
Person 2					
Person 3					

Person	Head squares	Body squares	Arm squares	Leg squares	Total squares
Person 1	1	3	2	2	**8**
Person 2	1	4	4	4	**13**
Person 3	1	5	6	6	**18**
Person 4	1	6	8	8	**23**
Person 5	1	7	10	10	**28**
Person 6	1	8	12	12	**33**
Person 7	1	9	14	14	**38**
Person 8	1	10	16	16	**43**
Person 9	1	11	18	18	**48**
Person 10	1	12	20	20	**53**

- *What patterns can you see?*
- *Are you surprised by anything you notice? Why / why not?*
- *Why does the total number of squares increase by 5 each time?*
- Each new person has 1 more body square, 2 more arm squares and 2 more leg squares.

5. Draw a diagram (2)

Additional resources required: squared paper and rulers (optional)

Slides 10 and 11

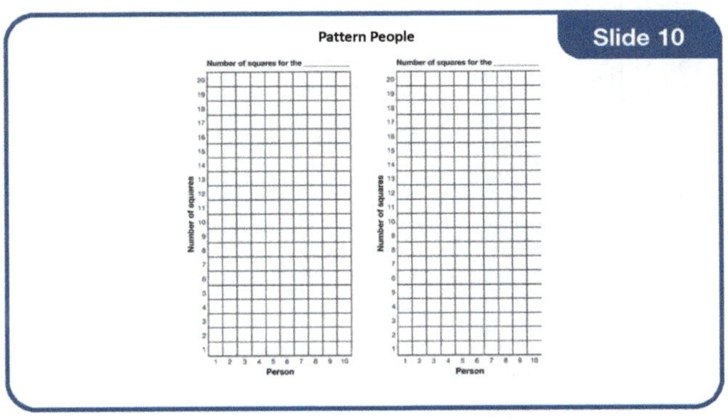

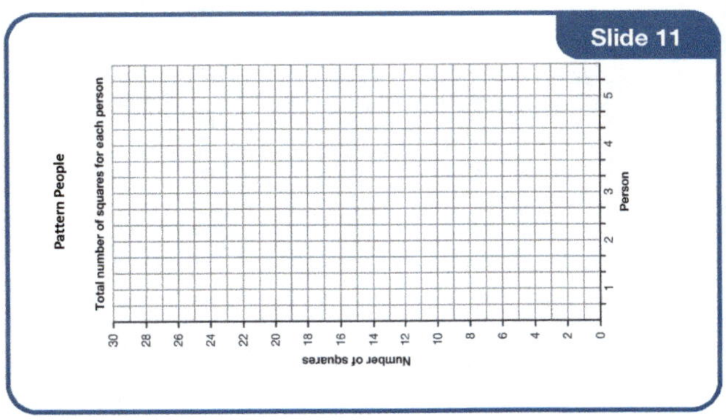

Print slides 10 and 11 and make them available for children to use if they wish.

Ask children to represent their data in a block diagram or bar chart. They could draw their graphs on squared paper, or use the block diagram template (slide 10) or bar chart template (slide 11).

Children may decide to graph the individual elements separately, to help them see exactly what is changing each time and what is staying constant.

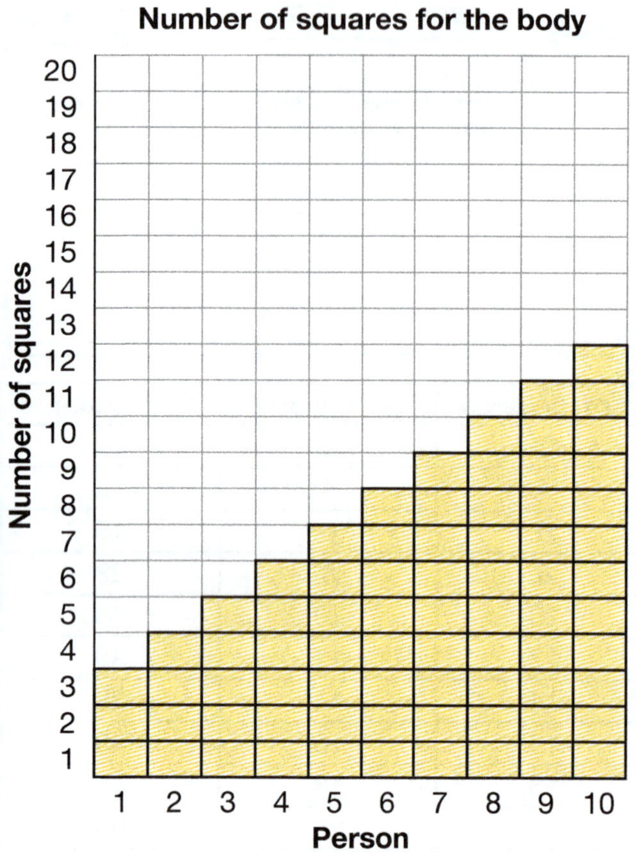

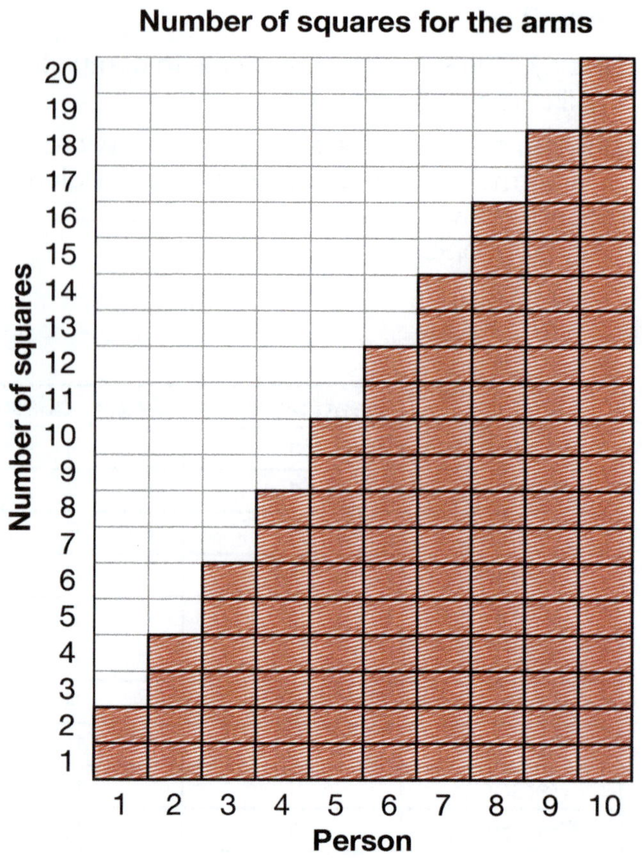

Block diagrams showing the number of squares for the body and the arms for the first ten people

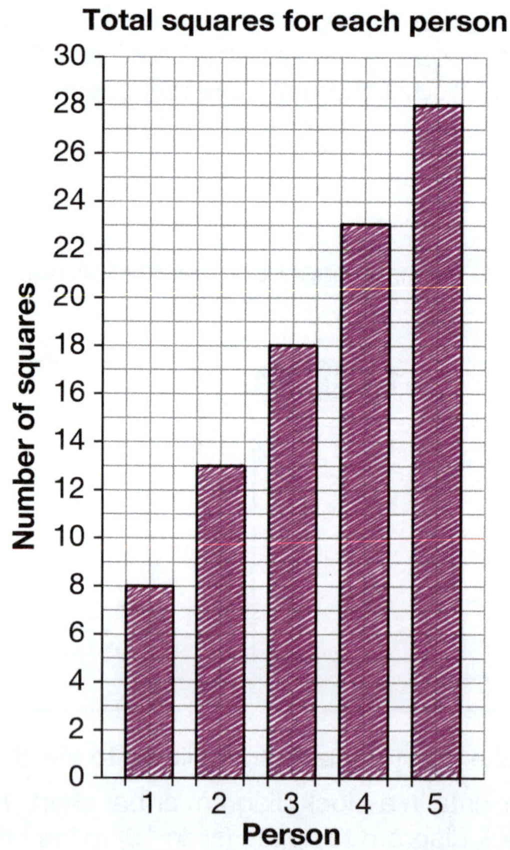

Bar chart showing the total number of squares for the first five people

Connect

- *What do your graphs show?*
- I can see that the number of squares needed for the head does not change. It is always 1.
- I can see that the number of squares needed for the body goes up by 1 each time.
- *How many more than the 'person' number is the number of squares needed for the body?*
- I can see that the number of squares needed for the body is 2 more than the 'person' number.
- I can see that the number of cubes needed for the arms / legs goes up by 2 each time.
- *How many more than the 'person' number is the number of squares needed for the arms?*
- I can see that the number of squares needed for the arms is double the 'person' number.

Collaborate

Refer to the **Collaborate** task of the **Pupil Resource**.

This task provides an opportunity for children to work in pairs or in small groups to create their models or diagrams.

Each child could build a different model (for example, one child builds person 1, another builds person 2, and so on) and then they could compare.

- *What's the same about each of your models?*
- *What's different?*

Alternatively, one child could create person 1, then pass their model or diagram along for the next child to add to it to turn it into person 2, and so on.

- *How many squares did you need to add?*

Recording of results could take place in a group table or individual tables.

Communicate

Refer to the **Communicate** task of the **Pupil Resource**.

Slide 12

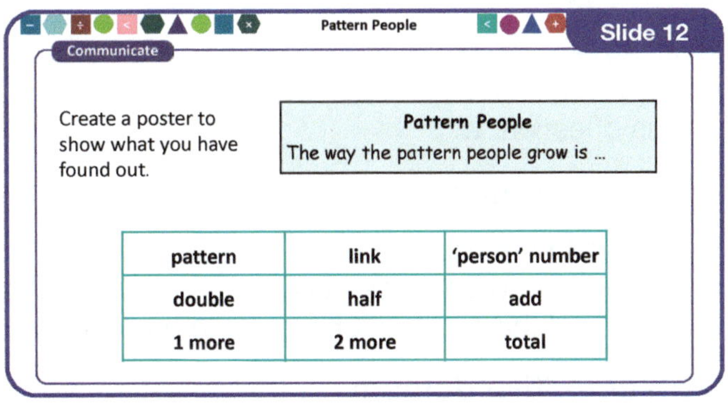

Display slide 12.

This task provides an opportunity for the children to create a poster that shows how the pattern people grow. This could include diagrams (or photos of models) with labelled explanations, tables and graphs.

The tables and graphs could be annotated to show where relationships and patterns can be found.

Each group could then present their poster to the class or another group.

pattern	**link**	**'person' number**
double	**half**	**add**
1 more	**2 more**	**total**

Create

Refer to the **Create** task of the **Pupil Resource**.

Slide 13

Display slide 13.

- *Were there any questions asked at the start of the task that you now want to think about?*
- *Is there anything new that you are wondering now?*

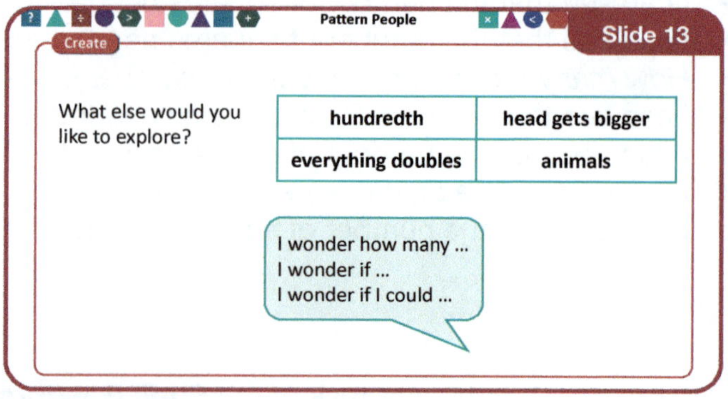

Children can use the word bank to prompt further questions of their own.

hundredth	head gets bigger	everything doubles	animals

- I wonder how many squares the hundredth person would be made from.
- I wonder what would happen if the head got bigger each time.
- I wonder if I could create a person where the arms, legs and body double each time.
- I wonder if I could make a growing pattern of animals made out of squares.

Consider

Refer to the **Consider** task of the **Pupil Resource**.

Slide 14

Display slide 14.

Encourage children to think about the strategies that they have used. The prompts below may help get children started.

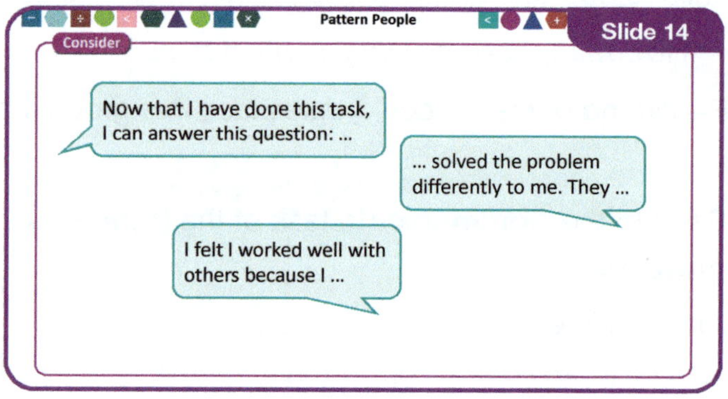

Cognitive learning domain

- I learned …
- Now that I have done this task, I can answer this question: …

Affective domain

- … was fun because …
- I said ‘aha’ when … because …
- … solved the problem differently to me. They …

Metacognitive domain

- I felt I worked well with others because I …
- Next time I work with a group, I will …